Healing Roots

Series: Epistemologies of Healing
General Editors: David Parkin, Fellow of All Souls College, University of
Oxford and Elisabeth Hsu, Professor of Anthropology, University of Oxford

This series publishes monographs and edited volumes on indigenous (so-called
traditional) medical knowledge and practice, alternative and complementary
medicine, and ethnobiological studies that relate to health and illness. The
emphasis of the series is on the way indigenous epistemologies inform healing,
against a background of comparison with other practices, and in recognition of
the fluidity between them.

Healing Roots
Anthropology in Life and Medicine

Julie Laplante

berghahn
NEW YORK · OXFORD
www.berghahnbooks.com

Published in 2015 by

Berghahn Books

www.berghahnbooks.com

©2015, 2018 Julie Laplante
First paperback edition published in 2018

Library of Congress Cataloging-in-Publication Data

Laplante, Julie, author.
    Healing roots : anthropology in life and medicine / Julie Laplante.
        pages cm. — (Epistemologies of healing ; v. 15)
    Includes bibliographical references.
    ISBN 978-1-78238-554-7 (hbk : alk. paper) — ISBN 978-1-78920-059-1
(pbk : alk. paper) — ISBN 978-1-78238-555-4 (ebook)
    1. Artemisia afra—Therapeutic use—South Africa. 2. Medicinal plants—
South Africa. 3. Traditional medicine—South Africa. 4. Medical
anthropology—South Africa. 5. Ethnopharmacology—South Africa.
6. Molecular biology—Social aspects. 7. Drugs—Testing. 8. Tuberculosis—
Chemotherapy. I. Title. II. Series: Epistemologies of healing ; v. 15.
    RM666.A775L36 2015
    306.4610968—dc23

                                                        2014029572

British Library Cataloguing in Publication Data

A catalogue record for this book is available from the British Library

ISBN: 978-1-78238-554-7 hardback
ISBN: 978-1-78920-059-1 paperback
ISBN: 978-1-78238-555-4 ebook

To Lo and Oiy for doing healing in life

Contents

✤ Illustrations

♣ Acknowledgements

Following *umhlonyane* on its trails and trials was done as a senior research fellow within the Biomedicine in Africa research group at the Max Planck Institute für etnologische forshung in Halle (Saale), Germany, led by Richard Rottenburg. I am very grateful to the institution for its full financial support of the research conducted from 2006 to 2010, as well as to the insights I received from colleagues throughout its unfolding. While conceiving *Healing Roots* I benefited from conversations with Trevor Pinch, Margaret Lock, Sheila Jasanoff, Stacey Langwick, Hans-Jörg Rheinberger and Vihn-Kim Nguyen, as well as research associates Wenzel Geissler and Ruth Prince. I particularly thank my research colleagues Virginie Tallio, Thamar Klein, Babette Mueller-Rohstroh, René Gerrets and Julia Zenker for their support and team spirit, as well as extremely helpful and insightful students Norman Schräpel and Thomas Thadevaldt, both in Germany and in South Africa. And I, of course, thank Richard Rottenburg for his thorough insights and support orienting the unravelling of the research as it came into being. The transnational research consortium conducting the preclinical of *Artemisia afra* followed throughout the research is The International Center for Indigenous Phytotherapy Studies (TICIPS), based in both the United States and South Africa, and I thank its participants for their collaboration and hospitality, in particular researchers from the University of the Western Cape. I here thank Quinton Johnson, Wilfred Mabusela, James Syce, James Mukinda and anthropologist Diana Gibson for their collaboration, time and knowledge. From the Institute of Infectious Disease and Molecular Medicine in Cape Town, I thank Dr. Muazzam Jacob, Alex Valentine, Willem Hanekom and Hassan Mohammed, as well as Nceba Gqaleni from the Nelson R. Mandela School of Medicine in Durban, Leszek Vincent and William Folk from Missouri University, Mark Estes from the University of Texas

Medical Branch, Wendy Applequist from the William Brown Center for Plant Genetic Resources at the Missouri Botanical Garden, Gilbert Matsabisa from the Indigenous Knowledge Systems branch of the Medical Research Council in South Africa, and Josephine P. Briggs, director of the National Center for Complementary and Alternative Medicine of the National Institutes of Health in Washington, D.C. From the 'open air', I give my deepest gratitude to Fire, Reuben and his family as well as to Phillip Kubukeli for their time and conscientiousness, as well as to Mrs. Skaap, Miranda, Thozama, Gustavo, Gerald Hansford, Nigel Gericke, Keith Wiseman, Jo Thobeka Wreford and colleagues such as anthropologists Leslie Green, Steven Robins and psychologist Lauren Muller for their time, thorough insights, hospitality and inspiration. I further thank Judith Farquhar, Veenas Das and Rene Devisch for orienting me towards more grounded and phenomenological approaches in anthropology at a moment when I really needed direction. For the coming together of this long journey into written form I thank Tim Ingold for giving me inspiration and guidance to move deeper into this approach to anthropology, a breath of fresh air I hope to bring to the book. I also thank all the wonderful students who have not only borne with me in moving through the new anthropological approaches I apply throughout this book, yet have pushed me further into them. For initial revisions I thank my parents Monique and Jacques Laplante, my colleague Scott Simon and my student Angela Plant for both their philosophical and linguistic insights, bringing the final touches to the first version of the manuscript. For final revisions I thank anonymous peer reviewers as well as all the wonderful people along the production line at Berghahn Books, in particular Caitlin Mahon for her meticulous care in 'the latest phase of the manuscript's voyage from computer to printed volume' (her lovely words). Finally, the person I should perhaps thank the most is Elisabeth Hsu, who truly shook up the first version of the manuscript, making sure it remained fully grounded as well as alive throughout. This is where I wanted to go, but I needed that final push to reach the desired direction. Thank you so much.

My love and gratitude to Mark and our children Oscar and Marlie, who have travelled and lived with me on two different continents, making all of this possible and enjoyable, as well as for opening my research to new artistic sensitivities.

✤ Abbreviations

A. afra	*Artemisia afra*
CAM	complementary and alternative medicine
FDA	Food and Drug Association
IIDMM	Institute of Infectious Disease and Molecular Medicine
IK	indigenous knowledge
IKS	Indigenous Knowledge Systems
MRC	Medical Research Council
MU	Missouri University
M. tuberculosis	*Mycobacterium tuberculosis*
NCCAM	National Center for Complementary and Alternative Medicine
NIH	National Institutes of Health
RCT	randomized clinical trial
SAHSMI	South African Herbal Science and Medicine Institute
S. frutescens	*Sutherlandia frutescens*
TB	tuberculosis
TICIPS	The International Center for Indigenous Phytotherapy Studies
TM	traditional medicine
UCT	University of Cape Town
UWC	University of the Western Cape
WHO	World Health Organization

Introduction

Tracing the Preclinical Trial of an Indigenous Plant

> Life, in short, is a movement of opening, not of closure.
>
> – Tim Ingold, *Being Alive*

Anthropology in life is an effort to 'follow what is going on' (Ingold 2011: 14). Life as a movement of opening can be understood in positive terms of healing, renewal, growth and paths along which life can keep on going; however, movement is also about falling down, destruction or dying, which are also paths of becoming something else. In such movements in life involving materials, some promise the riddance of malignant spirits or current bothersome 'body-in-the-world' orientations. Other movements, for their part, promise the elimination, destruction or slowing down of the replication of targeted bacterium. A nuance between the two movements may rely on a controlled study holding promise of defining a determined pathway that should apply to all biological bodies, notwithstanding the inhabited world, while the other embraces the indeterminacies in the moment, with a promise of the possibility of enhancing skills to deal with life's intricacies. In this book, life is left open-ended in following both humans and nonhumans, namely, plants, as they interlace to find mutually beneficial ways to heal within life-making processes in the inhabited world. Therein lies the main argument I put forth in this work regarding the question of 'knowing' medicine in healing; it is always being done and undone in life rather than something 'out there' to discover. I aim to show how this takes place as a 'wild' indigenous plant makes its way into human lives as a medicine, rather than as a weed or as food, for instance. More particularly, I trace how an in-

1

digenous plant is made to appear within a preclinical trial to test its efficacy against the tuberculosis pandemic as declared by the World Health Organization (WHO) in Geneva, Switzerland. The preclinical study of *Artemisia afra* against tuberculosis was undertaken in 2005 by a research consortium named The International Center for Indigenous Phytotherapy Studies (TICIPS, pronounced *tea-sips*), a trial financed by the National Center for Complementary and Alternative Medicine (NCCAM), a division of the National Institutes of Health (NIH) in Washington, D.C., in the United States and taking place in Cape Town, South Africa. To find how medicine is made to 'work' in healing through this process and at its edges, I thus follow a wild indigenous bush in its trails and trials of becoming a biopharmaceutical.

The bush is indigenous to the sub-Saharan region and is named *umhlonyane* in Xhosa and Zulu, *lengana* in Tswana, *zengana* in Sotho, *wilde-als* in Afrikaans, and wild wormwood in English (van Wyk and Gericke 2007: 142). Its Latin name in the scientific botanical nomenclature is *Artemisia afra* (Jacq. ex Willd). It is a multistemmed perennial shrub with grayish-green feathery leaves of the family Asteraceae (Mukinda 2005: 1), its areal parts having the strong characteristic odour of wormwood and a bitter taste (ibid.: 22). The genus name *Artemisia* honours Artemis, the Greek goddess of hunting (W. Jackson 1990) and protector of the forest and children. The specific epithet *afra* means from Africa. *Umhlonyane* grows 'wild' in the sub-Saharan region, as far north as Ethiopia, in particular along humid coasts, along streams and on the margins of forests at altitudes varying between 20 and 2,440 metres. It grows in bushes that can reach up to two metres in height, blossoming in the late summer, producing abundant bracts of butter-coloured flowers, each approximately three to five millimetres in diameter. *Izangoma*,[1] Xhosa 'diviner-healers who achieve their diagnosis and remedies through communication with ancestral spirits' (Wreford 2008), set it under bedsheets in preparation for healing. *Izinyanga* (specialists in herbal medicines, including Rastafarian *bossiedoktors*[2]) use this plant regularly, whether collected in the 'wild' or from a shoot transplanted in their backyards. '*Umhlonyane* is one of the oldest and best-known of all the indigenous medicines in southern Africa, and has such diverse and multiple uses that it should be considered a significant tonic in its own right' (van Wyk and Gericke 2007: 142). It is one of the most documented indigenous plants in sub-Saharan Africa, with the first accounts beginning in 1908; the list of uses of the plant in South Africa covers a wide range of ailments from coughs, colds, fever, loss of appetite, colic, headaches, earache and intestinal worms to malaria, diabetes and influenza; it can be taken as enemas, poultices, infusions, body washes, lotions, smoked, snuffed or drunk as a tea (van Wyk and Gericke 2007: 142). It has only recently undergone preparation for a randomized clinical trial (RCT), science's current gold standard[3], to test the efficacy of biopharmaceuticals, in this case against *Mycobacterium tuberculosis*.

The beauty of the preclinical moment is that the medicine is not yet made; it is the very process of making an object of study, in this case that will seek its orientation within *muthi*[4] multiplicities yet at the same time answer to a world pandemic in the One World–One Medicine–One Health[5] framework, that is, a process of closure to indigenous medicine conducted to gather evidence justifying a clinical trial. How these movements are reconciled (or not) in practice is thus what is of interest. A combination of motions such as a worldwide revitalized interest in wild plants; new approaches in molecular biology; the utmost commonality of *A. afra*'s medicinal use throughout sub-Saharan Africa; the 'African Renaissance', seeking pride in African solutions for its nation and for the world; HIV/AIDS driving a resurgence of tuberculosis; drug resistance to old drugs (the newest drug used to treat tuberculosis is more than thirty years old) spreading out of control (Check 2007); and the success of the isolation of artemisinin (*qing hao su*) of *Artemisia annua* L. (Chinese: 青蒿; pinyin: *qīnghāo*) in becoming a global health medicine against malaria (see Hsu 2010) may all partake in the coming to life of a preclinical study of *A. afra* phytotherapies in the treatment of *M. tuberculosis*. The biochemical, pharmacological and immunological activities of *A. afra* were studied in both Africa and America by TICIPS from 2005 until 2010. Tracing this specific preclinical trial in my study has led to previous toxicity trials, to parallel preclinical trials and to diverse and often conflicting knowledge practices within the sciences as well as without, including regulatory frameworks, politics and indigenous healing practices. Following the process of the preparation for an RCT, following the roots and routes, trails and trials, of *A. afra* becoming a biopharmaceutical (or not) through this initiative, constitutes the red thread, or macramé, tying the book together, bringing people, plants and the world into conversation as they endeavour to make medicine meaningful in healing.

This book stems from a long journey trying to understand how we come to 'know' medicine. In pursuing this quest along various paths, I came to an understanding of the RCT as an anonymous beholder of truth about the efficacy of medicine. The industry of clinical trials shows some of the greatest investments in the world. There are more than 50,000 clinical trials worldwide (Petryna 2007a). The NIH website alone shows more than 5,800 entries for clinical trials against HIV/AIDS and 2,700 against tuberculosis.[6] RCTs of indigenous medicine are just a very small piece of this industry, yet they are also increasing in scope today for a variety of reasons linked with those expressed above. In their offshoring, RCTs have become new spaces of care, healing and politics as well as new spaces of concern for anthropologists.

A growing body of research on clinical trials in the social sciences has attended to their histories (Marks 1999), ways the gold standard creates new worlds of treatment (Timmermans and Berg 2003), and how they create new politics of inclusion (Epstein 2007); others attend to the limits of these designs (Cartwright 2007) and their ethical variability (Petryna 2005, 2009); most take for granted, however, that there is a model 'out there' to begin with. A new 'praxiological turn', however, enables us to assess these practices as they are being done. A few researchers lead the way in this turn, namely, Mol (2002), in proposing an ontology of a multiple object, and Brives (2013), with a particular interest in following how participants 'do the body' within clinical trials. My work moves in this line of enquiry; however, the participants in my case study include the scientists and healers 'doing medicine'. Further, I follow a preclinical trial, an open-ended moment in which objects are not yet made nor closed; the RCT is but imagined to find ways to bring such closure. I am unaware of any other study done in this upstream moment of the clinical trial, let alone done in the preclinical trial of an indigenous medicine, a process that also remains understudied. Gibson (2011) studied the process of *Sutherlandia frutenscens* to be tested through a randomized placebo controlled clinical trial against HIV/Aids in South Africa, namely a trial in which TICIPS' researchers are also implicated. She attends to this trial as a messy, contested and ambiguous process akin to the approach I propose, showing how the indigenous plant is multiple on the grounds while it is relationally performed as a botanical plant entity in the scientific literature and clinical practice. She however does not attend to the way this is done in preclinical studies. Adams (2002; Adams et al. 2005) provides an anthropological outlook of the challenges in the clinical trial of indigenous Tibetan medicine, yet the process of the trial coming into being is not dealt with. She shows in part how the process implies the creation of a whole infrastructure as well as deals with the unfairness of the game for indigenous practitioners, but does not investigate the ways in which these negotiations unravel upstream. To the contrary, in my case study, the RCT has not come into being (if it ever will be), and I thus solely address how it is imagined and the ways the preclinical process unfolds.

Being in the open-ended moment of an imagined trial, one whose process is doubled by concerns with indigeneity[7], I largely move out of the controlled laboratory environments, since healers largely 'do medicine' in ways not taken up by scientists. While some scientists were clearly familiar with 'indigenous' practices that influenced their search in particular molecular configurations rather than others, they are, however, led to filter these out within their expertise, mainly bringing into being what they imagine will be recognizable within the RCT process, which in the end leaves very little traces of indigenous ways of doing medicine. This is revealing in itself; in a simultaneous quest to both open up prospects for new molecules and retrieve indigenous dignity[8], priority is clearly given to the former. In attending to indigenous medicine, the borders of the foreseen clinical trial requirements are, however, pulled in varying directions away and beyond the imagined model, namely, towards issues of indigenous dignity. The preclinical trial of an indigenous medicine thus shows different worlds of becoming along which medicine is being done in healing. In preparing for the trial, some of the histories of indigenous medicine, of clinical trials, of disease, of efficacies and of which kinds of life we aim to save or attend to are brought into being, while others are left to wither away.

The journey here told begins in August 2006 within the Biomedicine in Africa research group at the Max Planck Institute für etnologische forshung in Halle (Saale), Germany, wondering how biomedicine is both shaping and being shaped through its practices in Africa. My quest to find clinical trials with indigenous medicine in Africa was answered in November 2006 when four molecular biologists from TICIPS involved in the preclinical trial of *A. afra* replied in echo, sharing similar concerns as my own with issues of efficacy. The notion of efficacy was problematic in their dealings with *muthi*, seemingly conflicting with the narrowed-down notion of efficacy to which clinical studies attend, namely, one confined to controlled environments and precise physiological mechanisms of action. From that point onwards, I spent the next five years following the trails and trials of *A. afra* becoming a biopharmaceutical (or not). My fascination with the entanglements between indigenous and humanitarian medicine dates back to the early 1990s during research in the Brazilian Amazon (Laplante 2003, 2004, 2006, 2007). It progressively became a fascination with biopharmaceuticals, more precisely, with how they are made, taking me out of the jungle and into research centres, clinics and laboratories in downtown Montreal (Carrier et al. 2005; Laplante and Bruneau 2011). Following the trails of preparing an indigenous medicine to become a biopharmaceutical seemed to offer the possibility to reconcile some of the routes of knowledge in medicine that I had previously followed, bringing me full circle into new laboratories, clinics and laboratories in the United States and in South Africa, as well as back to the bush, mountains, valleys, townships and everyday lived experiences with indigenous medicine in South Africa.

It is these trails and trials of making medicine that I follow: from the laboratory to the bush and back, following how humans engage with a plant from its roots, indigenous to a people and to a place, to its possibility to travel in transnational One World–One Medicine–One Health initiatives. *A. afra* is not yet made into a biopharmaceutical (if it ever will be), and this is not as important as what I learned through the process; how the plant is transformed in its environment and beyond, who is involved, in which ways of knowing and not knowing life, which ontologies are brought into being, through which hopes and politics, and, perhaps most of all, what is left to wither away as irrelevant from the onset of the RCT process and what continuously emerges at its edges. The making of a desired tool to control a world pandemic is telling of the kinds of life attended to in humanitarian, clinical and scientific research. The negotiations, dialogues, actors, things, ancestors, gods, politics, standards, natures, cultures, bodies, molecules, animals and plants that are involved in this quest reveal larger worldly processes of this particular investment in knowing and not knowing medicine and, consequently, life. At stake is the encounter between, on one hand, perhaps the most standardized scientific research model in biomedicine and international ethical guidelines (the RCT) as followed by molecular biologists, immunologists, pharmacologists, plant systematists and ethicists and, on the other hand, one of the most sophisticated healing practices explicitly aiming to unstandardize and open ways for new orders, namely, those of South African Xhosa *izangoma* and Rastafarian *bossiedoktors*. The expertise of the indigenous healing practices are not here treated as exotica, yet rather as knowing practice, a notion earlier proposed by Farquhar (1994), which challenges and informs knowing practice of the scientific RCT community in various ways.

Knowing *as* movement (Ingold 2013: 1) is thus followed as it is being done and undone in making medicine to heal both within and without a preclinical process. To explain what is going on, I first introduce the plant and what has led up to its entry in the preclinical trial. Second, I find a way to follow what is being done and undone with the plant in practice: I delineate a phenomenological approach in anthropology enabling me to do so. Third, I follow these practices as I move through different places in which they emerge. I situate these traces in the worlds of indigeneity as undertaken by Rastafarian and Xhosa healers and in the One World–One Medicine–One Health path as undertaken by molecular biologists (chapters 3 to 7). I attend to ways these apparently separate routes weave themselves together in their known and foreseen efficacies as well as show where they diverge in the kinds of lives they attend to. What the trial could become, what its place in life could be, in the world that shapes us but that we shape as well is discussed in the conclusion in which I ask if ethics or aesthetics is the way forwards for anthropology.

I introduce *umhlonyane* as well as what I understand as 'knowing' in chapter 1. I begin to show some of the processes of making this 'wild' bush recognizable to science as it emerges in botanical inventories as *A. afra* and in publications with its particular concern with *M. tuberculosis*. What is done upstream in envisioning 'doing medicine' is explained as a process of making an object that will be able to filter through the imagined predesigned model 'out there', namely, the RCT. The preclinical process is found to be 'in the world', however, with an intention to withdraw both itself and the plant from the world, notwithstanding its assertion, in this case, to aim to recognize indigenous knowledge. Because of this announced dual program, which primarily attends to the prospect of a new molecule fighting *M. tuberculosis*, thus leaving behind the attention that might otherwise be given to ways of knowing *umhlonyane* in practice, chapter 2 delineates an anthropological approach that will enable me to attend to both aspects, notwithstanding the imagined RCT.

The anthropological approach I delineate in chapter 2 is precisely an attempt to see what it would look like to inhabit the preclinical trial without giving up life in the plant that is being tested, and in this way recognize indigenous ways of healing rather then set them aside as 'bias', 'placebo', 'culture' and even as 'indigenous knowledge'. Contrary to the current preclinical moment, which is about making an object, even assuming there is such an object 'out there' to discover, I thus align with *muthi* multiplicities, which leave this process open-ended and always in the making. 'Medicine' itself is thus not assumed to be unified and instead is seen as continuously emerging through new engagements with humans. The way the preclinical trial is done in practice is also continuously improvised, notwithstanding the imagined RCT, and even because of the imagined RCT, which often requires even more ingenuity to fit within the context at hand. The way I access the emerging ontologies of practice both inside and outside the preclinical process is through a phenomenological approach in anthropology, one done through fine-tuning my attention to the ways 'medicine' is being done and undone in everyday practices. I trace medicine anthropologically, both as method and theory. This is a process attuned with the tools developed by anthropologists in order to maximize my ability to grasp what is going on, one which I delineate through intertwined notions of embodiment, sentience and medium, which relate to the 'worlds of becoming' I aim to understand.

Chapter 3 tells what happened in the preclinical trial of *A. afra* as I immersed myself in and around its practices. I delve into the experience of the complexities of *muthi* and the preclinical trial 'in life', or in the way they have become within my grasp in the process of research. I tell of some of the *A. afra* trails and trials I followed, how I engaged with some of the actors and mediums involved in the preclinical trial, and what it felt like to move through these mediums.

I move from laboratories to botanical gardens, backyards, markets, shops, valleys, farms and festivals. I tell how I entered the preclinical trial with molecular biologists who then pointed towards further trails to follow with other scientists, as well as, eventually, with Rastafarian *bossiedoktors* and Xhosa *izangoma*. I describe how knowing *umhlonyane* appears in their respective practices and everyday lives. Moving from laboratories to backyards offers some thoughts on the importance of propinquity in indigenous practices, which contrast with the kinds of practices done in controlled environments, showing part of the dissonance between their intentions in doing medicine.

In chapter 4 I move into *muthi* as part of South African national politics, showing how the biomolecular line of the trial pulls towards questions of indigenous knowledge and human dignity. The notion of 'indigenous medicine' is discussed as it emerged in anthropology, in 'world health' and in the South African context, consolidating into institutions such as the NCCAM branch of the NIH in Washington, D.C., and into the Indigenous Knowledge System (IKS) branch of the South African Medicine Research Council (MRC). The colonial histories of *muthi* in South Africa are foreshadowed with relation to ways they emerge within current practices, namely, within those of Rastafarian *bossiedoktors* visited in-depth in chapter 5 and those of Xhosa *izangoma* described in chapter 6. The two latter chapters explain the ways I became acquainted with these healers, becoming 'part of the struggle' with Rastafarian *bossiedoktors* and remaining 'part of the problem' with Xhosa *izangoma*. Moving deeper into ways of knowing in-the-world, I show the ways *wilde-als* is part of Rastafarian *bossiedoktor* practices and how *umhlonyane* (dis)appears within Xhosa *isangoma* practices in the Cape. In both cases I show how their current practices unravel in the everyday, explaining the traditions they evoke and enact when expressing their healing efficacies 'in the inhabited world'. In particular, 'sounds' are shown to be central to both become a healer as well as be healed, pointing towards deepened engagements in-the-world, to the multiplicity of medicine and to particular kinds of lives that are attended to. From this immersion into indigenous ways of healing, I move into molecular biologists' practices as they unravel within and without an imagined clinical trial.

In chapter 7, I explain how molecular biologists pull the preclinical trial towards concerns with indigeneity while at the same time imagining new clinical trials. I explain life in molecules as a movement of opening, showing correspondence with Xhosa *isangoma* practices; however, this movement closes again as it aims to fit the RCT protocols, thus losing its dealings with indigenous medicine. As such, innovation is exclusively attributed to science, which thus becomes a path undertaken for political reasons as well as scientific ones. I suggest thinking of practices in terms of improvisation can be a way to surpass this impasse. Finally, I briefly discuss how Rastafarian *bossiedoktors*'s explicit opposition to imperialism, and thus implicitly to the One Health Initiative currently claiming this positioning, tends to the very matters the One Health Initiative lets wither away, expressed as 'One Love'. The conclusion attends to hopes moving through the preclinical trial in imagining what the clinical trial can achieve. The question of the efficacy of A. *afra* in this study is found to be woven into lines or overlapping motions of hope in particular kinds of life that translate themselves along the thrust of humanitarian efforts to 'save lives', of the 'African Renaissance' and of 'indigeneity'. Attuning attention to knowing aesthetically within these life flows is proposed as a way forward for anthropology.

Notes

1. *Izangoma* is the Zulu plural form of *isangoma*. In the same logic that 'z' means plural, the term *izinyanga* is the Zulu plural form of *inyanga*.
2. Afrikaans term for bush doctor.
3. In its original sense, used by economists, the gold standard referred to a monetary standard under which the basic unit of currency was defined by a stated quantity of gold (Claassen 2005: 1121). The establishment of a gold standard was used to compare the value of currency. When the term moved into medicine, it referred to the current best tool upon which the value of other tools could be compared (Rudd 1979: 627). Some scientists immediately rebelled against the use of the term 'gold standard' for clinical procedures. A biochemist, for instance, found it presumptuous for a biological test, 'because the subject is in a state of perpetual evolution (and) gold standards are by definition never "reached" (Duggan 1992: 1568). A test is never "perfect" and always ready to be replaced by a better research tool (Claassen 2005: 1121). The term nevertheless did progressively come to denote 'the best standard in the medical world' and beyond. The RCT has become 'one of the simplest but most powerful tools of research' (Stolberg et al. 2004: 1539). Nonetheless, the RCT has lost its sense of simply being a reference upon which to compare other research methods and rather has

become a dogma that implies perfection or a way to access definite 'truth' about the efficacy of a medicine.

4. *Umuthi* in Zulu means 'tree' or 'bark' and also 'medicine'; the word is rendered as *muti* or *muthi*. Most South Africans use the term *muthi* to refer loosely to all forms of medicine that are not obviously biomedical, but also know that the word itself has deeper connotations. I intend to use the term to refer to ways of engaging with nonhuman life in healing that are not obviously biomedical and will bring subtleties to my understanding of *muthi* in chapter 4.

5. One World–One Medicine–One Health, or the One Health Initiative which emerged in the years 2000, is a movement and worldwide strategy for expanding interdisciplinary collaborations and communications in all aspects of health care for humans, animals and the environment. It joins efforts, amongst others, of the WHO, the United Nation's Food and Agriculture Organization (FAO) and the World Organisation for Animal Health (OIE) in attempts to ameliorate health care for the twenty-first century and beyond. One Health is portrayed as an umbrella encompassing varying scientific expertise in health, somewhat re-enacting the modernist dream of 'health for all' led by the WHO, similarly excluding any form of healing which might work through other routes than the current privileged molecular biological route endorsed by biomedicine. While clinical trials have emerged in the medical sciences during modernity in the 1950s, many instances show how the model has travelled through the disciplines and this may in part explain how a new umbrella is imagined to encompass them. As will become clearer in the conclusion, I argue that 'oneness' (or the umbrella) is an attempt to re-establish a universalist objective stance, a stance that places 'one nature' as the backdrop for multiple incommensurable cultures, only to attend to the first and thus excluding rather than including people, plants, animals and the lived environment within it's scope. See http://www.onehealthinitiative.com/ (accessed 20 August 2014).

6. See http://clinicaltrials.gov/ct2/results?term=clinical+trial+tb&Search=Search (accessed 4 November 2013).

7. 'Indigeneity' is a highly politicized term officialised with the United Nation's Declaration on the Rights of Indigenous People in September 2007 and to which I will return in depth in chapter 4. Also see Pelican (2009) on the complexities of indigeneity and autochthony.

8. The notion of indigenous dignity in South Africa can be traced to Steve Bantu Biko, the founder and martyr of the Black consciousness movement to free the people oppressed during colonial and apartheid regimes in the 1960s and 1970s. A Steve Biko Heritage Centre promoting self-reliance projects affirming that blacks can earn their own keep with dignity opened in 2012 in the Eastern Cape. Biko's legacy is today also taken up in a movement towards indigenous African bioethics (Behrens 2013: 32).

Chapter 1
Knowing *Umhlonyane/ Artemisia afra*

The common world (of what the universe is really made up) is known by the scientists, but invisible to the eyes of the common people. While what is visible, lived, felt, is, to be sure, subjectively essential but utterly inessential, since it is not how the universe is made up. This means that when the time comes to tackle the political work *par excellence,* namely the definition of what sort of world we have in common, scientists can say that the task is *already completed* since the primary qualities are all summed up in one Nature.

– Bruno Latour, 'When Things Strike Back'

Umhlonyane is easily recognized by its grayish-green foliage and its pungent, sweet smell as soon as it is cut, brushed against or even as it sways with the wind. When touched it leaves the fingers sticky. With its bitter taste and refreshing aroma, a recurrent memory evoked when I asked about *umhlonyane* is that it smells and tastes like 'medicine'. It accommodates perfectly well to the townships of Cape Town, where most of the healers I met live with one of its bushes growing in the backyard, and most of the homes I visited kept it dried in the kitchen, ready to make tea. Whether it is the plant's accessibility, flexibility and ability to make itself visible, labelled, loved and cared for by humans, or humans 'discovering' its properties as medicinal in 'nature', both routes contribute to making ways along which the plant's life keeps on going. *Umhlonyane* has successfully managed to partake in the everyday healing of both ordinary and expert people in the Cape, as well as progressively being made amenable to scientific experimentation. In this chapter I explore the latter route and will return at length to the ways *umhlonyane* emerges in everyday practices in the following chapters, since it is mostly left unattended in the preclinical trial process per se, notwithstanding the specific TICIPS trial's claim to be a way to recognize indigenous medicine. I here touch upon the little that is known of everyday uses and healing performances of *umhlonyane* for the purpose of preclinical trials to gather evidence justifying a clinical trial.

What is well attended in TICIPS' preclinical trial of *A. afra* is concern with the prospect of new molecules against *Mycobacterium tuberculosis*, giving clear priority to measurable molecular mechanisms as the ways of knowing the plant. The first section after this introduction nuances this understanding of 'knowing' to foreshadow both the problems arising when indigenous knowledge is to be recognized in the preclinical process and a way I propose to avoid reproducing these problems in my own anthropological account. Second, I address how the preclinical process currently unravels in the tradition of botanical inventories, documenting the properties and technical applications of the indigenous plant of interest. In the third section I show how a similar process is undergone with regard to the disease of interest, in order to reach an experimental form of tuberculosis upon which to try the plant. A fourth section describes how *Artemisia afra* is prepared and tested on an experimental form of tuberculosis in a safety study conducted by Mukinda and Syce (2007). In a fifth section, the results of the ongoing TICIPS preclinical trial of *A. afra* are discussed as published by Ntutela et al. (2009) and as explained through conversations I've had with some of the researchers involved. A final section examines the process of making objects amenable to the experiment in the laboratory, in view of undergoing the clinical process and what may already be lost and found in its preclinical phase.

Knowing

With 'knowing', I allude to attuned attention of the lived body-person in the world. Not knowing thus implies leaving skills of enhanced engagements in-the-world unattended. In doing so, I challenge a few commonly held assumptions about knowledge in Western scientific thought. First, I challenge the assumption that knowing is progressively accumulated through empirical studies in controlled laboratory environments. This I define as gathering information (giving form to the mind, or deforming it). Knowing I rather understand as continuously emerging movement done through engagements in-the-world. I thus question common statements that on one side recognize that the use of herbal medicines is widespread, yet on the other side state that 'very little is known about these medicinal plants' (Mukinda 2005: 12). In a sense, the latter statement is in line with my understanding of 'knowing'; it states that little is known *about* the plants, but omits that much is known *with* the plants. While most scientific accounts begin their assessment of a new aspect of *umhlonyane* by relying upon the lack of knowledge *about* the plant, they simultaneously rely upon the World Health Organization (WHO) survey indicating that about 70 to 80 per cent of the world's population relies on nonconventional medicine from herbal sources as their primary health care. What is clear is that there are useful engagements with plants in-the-world, yet these are not recognized as 'knowledge'; knowing is exclusively attributed to information gathered through scientific procedures, namely, experiments demonstrating a cause-and-effect relation between a part of the plant and the human physiological metabolism. I argue this information can become knowing; however, there is no guarantee this will occur in practice.

Experiment has become the highest mode of investigating life in positivist science, a way of knowing that has prospered from Claude Bernard's 'control experiment' in his *Introduction to the Study of Experimental Medicine* in 1865 and that has become familiar to generations of scientists thereafter (Pickering 2000: 13).

> In the last quarter of the nineteenth century, medicine experienced a scientific revolution. The revolution occurred not in the practice of medicine but in what came to be known as the medical sciences: bacteriology, physiology, physiological chemistry and pharmacology, each of which in turn acquired the means to produce and manipulate in the laboratory the phenomena of disease' (Marks 1987: 35).

Biologist Jacques Loeb (1859–1924) brought this ideal into practice and from the late nineteenth century onwards, many experimental biologists began to define their work around the control of organisms, envisioning manipulation, transformation and creation of all the phenomena subsumed under the word 'life' (Pauly 1987: 5). Clinical evaluation, no less than pre-clinical testing, thus came to be conducted according to the canons of external scientific experimentation. 'Controls' facilitate staging an experiment to observe the operation or mechanism of interest as well as become a means of gaining control over living processes (ibid.: 144). In this operationalist view of science – the ability to control was the measure of knowledge. This ability to control through demonstration of a procedure is still today the measure of knowledge in empirical practices as upheld in scientific accounts about plants; clinical trials make knowledge synonymous with the outcome of an experiment. As such, a clinical trial generates knowledge of precise biological mechanisms and pathways in a closed environment, at the price of losing sight of their interweaving in mediums.

Second, I challenge the corresponding assumption that knowing alludes exclusively to the sole ability to represent the modus operandi. Murray Last's 1981 seminal article alerts us to the importance of knowing about not knowing of both anthropologists and the people they study in imagining medical 'systems'. Littlewood's 2007 edited book takes Last's article as its point of departure. The introduction of the anthology attends to the need for anthropologists to not add coherence or suppose preexisting coherence, in particular with relation to health, healing and medicine. One reason not to assume preexistent knowledge by members of social groups and their healers is that 'in the extreme case, local treatments are seen as efficacious precisely because their mode of operation is mysterious or even unknown' (Littlewood 2007: ix). While I agree we should not add coherence or suppose preexisting coherence with what people know in medicine, I disagree with how Littlewood seems to equate 'not knowing' with a lack of interest in the modus operandi. Such a stance assumes there is a mode of operation 'out there' to be 'discovered' and that can be equated with knowledge. Knowing is here clearly measured in terms of empirical demonstration of biological mechanisms capable of being verified or disproved by observation in an experiment.

If, however, knowing is also the ability to engage successfully with herbal medicines for healing in everyday life, as I suggest, then knowing empirically in laboratory studies might not be so knowledgeable in this respect. Further, if empirical studies are done on an herbal medicine, it is precisely because very much is known about their benefits as used in the everyday as well as in sophisticated practices unravelling in healing sessions. Hence, I assume attuning attention to entanglements of people and things in-the-world is knowing. Encyclopaedic inventories document some of these practices together with results from laboratory studies showing a plant contains such and such chemical elements; this is information rather than knowledge and it only rarely translates back into everyday practices. Further, controlled studies may show the plant to have some bioactive mechanisms and pathways whose actions can be described in relation to a particular disease, also providing information rather than necessarily meaningful knowledge. In all cases, there has been attuned attention to one or another aspect, or to the entirety, of the plant either in controlled environments and/or in the 'open air'. Ways of knowing medicine are thus multiple and always being done through new engagements in-the-world.

This leads to a last point of clarification of my positioning with regard to empirical thought as it is currently made out to be the measure of knowledge. Testing theories against observations of the 'natural world' in empiricism relies upon a particular notion of perception that assumes the effects of stimuli can be measured upon a passive body-object. As such, biological bodies to be tested upon experimentally are made into a passive body-object simply awaiting the proper stimuli or molecules to heal, even if only temporarily. I here rather endorse Merleau-Ponty's (1945) phenomenology of perception, which assumes mind and body are always in continuous entanglements in their engagements in the world. In this notion of perception there are no longer objects and mental programs preceding action; these are continuously being done through new involvements in the world. Thus, continuous entanglements between seeing and moving, prior experiences of seeing or knowing, are always fed back through movement so that we are always engaging in the world or medium in new ways. In this notion of perception, it is the active engagement with the environment that elicits stimuli.

An experiment that assumes a passive human body (or 'intact' cell) upon which to test the safety or efficacy of a molecule (or a fixed dosage of plant) thus appears incomplete. This is because it disregards the ongoing movement of the body in the world as well as its previous experiences, which will take part in the synergies between the person and the plant preparation or molecule in the inhabited world. Assuming that previous experience is in constant entanglement with the ways the world is perceived (and in reaction with) in the moment makes it so that such an experiment to test a 'thing' upon an otherwise perceived 'passive' human (or 'intact' cell) no longer holds as the sole necessary path towards 'truth' of the 'efficacy' of an isolated bioactive compound. The proposed phenomenological stance in anthropology, which will be delineated in further detail in the next chapter, is deemed more appropriate, as it enables accounting for ways of knowing life through both the empirical route (which implies active engagement with the environment, even if this disappears in its results) as well as through lived bodily experiences as done in the everyday and through orchestrated performances. It is with this approach that I follow the ways ontologies are brought into being in practice in the preclinical trial.

It is in dealing with the entanglements of people and things in-the-environment as both natures and cultures that I situate my positioning. These engagements are not preprogrammed yet are learned through doing, attuning one's attention in the world, or what Ingold proposes we name 'skills' (2000) and later 'knowing from the inside' (2013). This heightened awareness, or attuned attention, is what I refer to as 'knowing', while 'not knowing' refers to what is of lessened interest and engagement (which can apply as much to not wanting to know a mode of operation as to not wanting to know a lived experience). This approach reverses or balances out some of the assumed hierarchies of knowledge as well as does not leave empirical knowledge intact; it is precisely the equation of knowledge with a demonstration of the 'mode of operation' as upheld in scientific research that it dislodges (at least from its position of exclusivity as 'knowing'). The proposed approach corresponds more closely to what occurs in the practices of the preclinical trial, as the scientists (like the healers) are not completely in line with the clinical model they aim to follow in one way or another.

The clinical model, the randomized clinical trial (RCT), was designed within a modernist project placing cultural diversity in a background of natural universality, a backdrop of 'one nature' in which diverse cultures play themselves out. Descola (2005) has named this ontology 'naturalist', as Husserl (1950: sec. 86) had already suggested; Western thought, or science in its broad dominant positivist stance, equates knowledge with the empirical knowledge of 'nature', while what is lived falls into 'culture' and, more precisely, into a category known as the 'placebo effect' in clinical studies. 'The point of this seemingly innocent divide is that it is a formidable political ploy' (Latour 2000: 118). It is the parts of nature whose mechanisms are measured, demonstrated and described that are commonly equated with 'knowing' (atoms, drugs, disease, molecules), while the lived and the felt in the world that entangle with materialities are left to wither away as secondary qualities. Keeping this hierarchy becomes especially problematic in the case of the trial of an indigenous medicine whose healers excel in tailoring and deepening those very lived and felt experiences in the world. Following a predesigned model within this ontological divide becomes a process of closure to indigenous medicine, since it disregards the lived and the felt, or the skills acquired in life. This in turn makes the process conflict with its own dual objective to simultaneously isolate a molecule (as required by the RCT standard) and recognize the dignity of a people (as in the recognition of indigenous medicine). How this is made to cohere (or not) in practices becomes of utmost interest for anthropological enquiry. I begin this endeavour by delineating some of the information gathered about *A. afra* as it was being made amenable to scientific experimentation. Along these lines, I introduce the plant as it enters human lives in the laboratory and as it is being prepared for clinical studies. This endeavour takes its first steps in the form of botanical inventories.

Tracking *A. afra*

> As early as the sixteenth century, travellers were being advised to observe indigenous practices and to collect material with a view to extending European *materia medica*.
>
> – Roy Ellen and Holly Harris, 'Introduction'

A. afra (figure 1.1) is one of the oldest and best-documented of all the indigenous medicines in southern Africa. Its reported healing properties and use in specific ailments can be found in a number of encyclopaedic inventories from as early as 1908. It is described as a popular bitter tonic and appetite stimulant in the Cape region of South Africa (Dykman 1908; Rood 1994; Thring and Weitz 2006). In his book *Zulu Medicine and Medicine-Men*, Bryant notes that its leaves are used as a general specific against febrile complaints:[1] 'a

Figure 1.1. *Artemisia afra* (Jacq. ex. Willd), December 2007

double handful of the leaves being infused as tea with a quart or so of hot water, and administered either as clyster or emetic' ([1909] 1966: 53). The infusion of the leaves and stems of *Lippia asperifolia* and *A. afra* are mentioned being used as a formula for fevers, influenza, measles and as a prophylactic against lung inflammations by Watt and Breyer-Brandwijk (1962). Decoctions of *Tetradenia riparia* and *A. afra* with salt are more specifically documented to be used to treat coughs in the Transkei region of the Eastern Cape of South Africa (Hutchings et al. 1996), attesting to combinations of plants for specific healing purposes. *A. afra* is also described as being used in the forms of enemas, body cleansing, lotions and infusions; it is smoked and sniffed (van Wyk and Gericke 2007: 142).

Other reports of the use of *A. afra* in botanicals state that it is taken in the form of steam or vapour; fresh leaf rolls used for inhalation purposes; an infusion filtrate used as a wash for haemorrhoids and in the bath to bring out the rash in measles, to soothe fevers, sores, rashes, bites and stings, and to wash wounds; and as an eye bath diluted with warm water to soothe red, smarting eyes (Roberts 1990). Leaves have been reported to be commonly smoked by some tribes to release phlegm and ease and soothe a sore throat and coughing at night (ibid.). In his thesis about the toxicity of *A. afra* in rodents, Mukinda (2005: 25) mentions a variety of further uses of the plant as found in botanicals:

> The Venda and Twana people are reported to make a solution (wash) of the plant for skin ailments, to draw out pimples and boils by applying them as a poultice. The warmed leaf is also used as an excellent and a soothing poultice over a painful neuralgia, mumps swellings and on a sprained or strained muscle, and bound over the stomach for babies with colic. In addition, the *A. afra* leaves may be rolled and inserted into the nostrils to alleviate headache and stuffy nose or a fresh leaf tip may be inserted and packed in the gaps of the teeth to relieve toothache.

The Zulu people are also reported to make an infusion of *A. afra* by grinding up the leaves and adding hot water, and then give this as an enema to children with worms and constipation (Roberts 1990; Watt and Breyer-Brandwijk 1962). A more recent botanical inventory aiming to bring more context with the use of plants explains how 'The Sotho people of South Africa make small plugs from the plant's soft, feathery leaves and insert them in their nostrils to clear their nasal passages. It has a very pleasant sweetish, lavender-like smell and a similar effect to inhaling vapour rubs such as Viks Vapo-Rub – a most popular brand in Africa' (Dugmore and van Wyk 2008: 57).

Overall, an exceptionally wide diversity of uses of *A. afra* have been recorded (Dykman 1908; Watt and Breyer-Brandwijk 1962; Roberts 1990; Iwu 1993; Hutchings et al. 1996; van Wyk et al. 1997; Dyson 1998; Neuwinger 2000; von Koenen 2001; Thring and Weitz 2006; van Wyk and Gericke 2007), including the treatment of colds, influenza, sore throat, asthma, pneumonia, blocked nose, stomach ailments, chills, dyspepsia, colic, croup, whooping cough, gout, flatulence, constipation, gastritis, poor appetite, heartburn, internal parasites, measles, headache, earache, gout, heart inflammation, rheumatism, diabetes, malaria, wounds, as a purgative, and the list continues, showing open-endedness rather than closure, nevertheless offering a panoply of documented historical use to follow up on in laboratory studies. N. Liu et al. (2009) conclude that these uses indicate that *A. afra* possesses antiviral, antibacterial and anti-inflammatory activities.

Scientific curiosity surrounding *A. afra* with an interest in assigning a chemical structure to its oils can be found in a first publication that appeared in 1922 (Goodson 1922). This interest in the therapeutic effects of the essential oils in the flowering tops of *A. afra* was pursued sporadically until 1988.[2] *A. afra,* however, really began interesting scientific researchers in the 2000s, and more particularly from 2005 onwards, with an average of eighteen publications per year (Patil et al. 2011: 2). In this second wave, the interests in the volatile oils (mainly 1,8-cineole, α-thujone, ß-thujone, camphor and borneol) in *A. afra* shifted towards interests in its flavonoids (apigenin, hesperetin, kaempferol, luteolin and quercetin).[3] Waithaka (2004) for instance conducted quality control studies of flavonoid-containing medicinal preparations, including preparations with *A. afra.* Muganga's (2005) toxicity study of luteolin levels in folkloric preparations and from *A. afra* aqueous extract in the vervet monkey as well as Mukinda's (2005) toxicity studies of luteolin from *A. afra* in mice and rats, also attest to increased interests in flavonoids. The scientific studies of extracts of *A. afra* are oriented towards its antifungal, antibacterial, antioxidant, anticancer, antimalarial and antitubercular activities, amongst others, all corresponding to contemporary declared global health concerns. The preclinical trial that I follow in this book is preoccupied with the antitubercular properties of the plant, in particular those that can attenuate *M. tuberculosis,* which has a history of its own in clinical studies. I will attend to this briefly before I return to *A. afra* as it converges with *M. tuberculosis* in scientific research.

Experimental Tuberculosis

Tuberculosis holds a reputation to have plagued humankind worldwide for thousands of years (Lawn and Zumla 2011). It is slow to replicate itself and has a capacity to persist for long periods of time in a latent state. Humans can thus cohabitate with the bacterium, as with numerous other bacteria, for their entire lives. Tuberculosis has multiple causes, processes and trajectories and is linked with lethargy, poverty and malnutrition, and today with HIV/AIDS, which weakens the immune system. *M. tuberculosis* is likely the most common in humans, although Bowker and Star (2000: 171) warn not to forget *Mycobacterium bovis* and *Mycobacterium avium,* which are also common. While it is most often linked with the lungs and coughing, tuberculosis does not appear in a single place in the body; it can affect the lungs, but lesions can also appear in other organs and tissues. As it tends to spread through the body, Star (1989) notes the tendency to implicate tuberculosis in all investigations of nervous and brain disease in the nineteenth century. 'And indeed even pulmonary tuberculosis – its most common form and one of the greatest killers in the history of humanity – cannot be simply classified' (Bowker and Star 2000: 172). More recent 'advances in mycobacterial genomics are now providing evidence that the amount of sequence variation in the *M tuberculosis* genome might have been underestimated and that some genetic diversity does have important phenotypic consequences' (Lawn and Zumla 2011: 60; see also Smith et al. 2009; Gagneux and Small 2007).

The disease is not only multiple; it also moves with treatments, bringing new complexities to the forefront, showing it is a continuous 'work in progress'. It lives in humans for years before any symptoms emerge, with a capacity to defy attempts to cure it; during an interview with a TICIPS' immunologist, he mentioned tuberculosis was thus somewhat 'intelligent'. As with *A. afra* becoming noticeable to the scientific eye through a process of documentation of its uses in terms recognizable within the laboratory, such is the case with bacteria. Similarly, to 'know tuberculosis' is equated to extracting it from its contexts and successfully identifying and describing the bacillus in the laboratory. The 'discovery' of *M. tuberculosis* along this particular route is attributed to the German scientist Robert Koch in 1882. The disease had been formerly named 'consumption' and thought to be caused by small living creatures that were transmitted through the air to other patients (Marten 1720). Bacteria, as we call them today, are considered to be among the first life-forms on earth. They inhabit soil, water, people, animal and plants. It was in 1944 that an antibiotic cure called streptomycin was found in soil bacteria, and this remains the basis for most medicines against tuberculosis we use today.

The first randomized curative trial for tuberculosis, both double-blind and placebo-controlled, was done with streptomycin in 1948.[4] The history of chemotherapeutic trials is reputed to be filled with errors due to empirical evaluation of drugs (Hart 1946, in Streptomycin in Tuberculosis Trials Committee 1948: 769): '[t]he natural course of pulmonary tuberculosis is in fact so variable and unpredictable that evidence of improvement or cure following the use of a new drug in a few cases cannot be accepted as proof of the effect of that drug' (ibid.: 769). To counter the complexity of the course of the disease, a rigorously planned investigation with concurrent controls was designed. The solution was in fact the controlled double-blind trial, and as such the method still holds to be the best to deal with this multiplicity of disease; namely, it deals with multiplicity by disregarding it and focusing on a single experimental form (as also done with multiplicity in *A. afra*). Controls are necessary since no two patients have an identical form of the disease, and this way as many of the obvious variations as possible are eliminated.

In the 1948 Medical Research Council investigation of streptomycin in tuberculosis, the type of case to be investigated was narrowed down as follows: 'acute progressive bilateral pulmonary tuberculosis of presumably recent origin, bacteriologically proved, unsuitable for collapse therapy, age group 15 to 25 (later extended to 30)' (ibid.: 770). The trial featured blind assessment (a double-blind trial, where both researchers and patients do not know who receives the treatment or the placebo), one of the core features of current RCTs. The golden era of RCTs is modernity,[5] and they have only slowly been put under scrutiny during the past decades. The most reviewed aspect of RCTs is precisely their attempt to eliminate 'bias', which implicates the placebo control (or blind assessment), random assignment to comparable groups and inferential statistics as a surrogate for determinism (Kaptchuk 1998: 1722), all which are crucial to reassess within a trial aiming to recognize indigenous medicine, as the following chapters will show. It is, however, within such parameters, aiming to exclude the placebo, that the disease is delineated and tested upon.

With more information accumulated about the mechanisms of the progression of tuberculosis, the disease of tuberculosis seems to have gained unity through time and attained the status of object, perhaps as it moves away from its lived experience in the world. When looking at large populations over time, it can seem homogenous and remain a useful category: 'the further away one stands from the disease of tuberculosis, the more it appears to be a single, uniform phenomenon' (Bowker and Star 2000: 165). However, even at a distance, tuberculosis 'has historically proved an elusive thing to classify' (ibid.: 165). Upon moving closer into the experience of the disease, its borders often fade away. As it spreads over long periods of time, tuberculosis moves through the body as well as changes form, making it difficult to correspond to specific trajectories of the lived experience of its symptoms within a context that will have some part in the course of the events, whether it be through treatments, family, friends, livelihood or one's own ways of engaging with these realities. The practicality of classifying tuberculosis allows officials in public health to act upon it with a treatment and a trajectory (ibid.: 175).

When streptomycin can be said to have efficacy against tuberculosis following such a precise experiment, it would be more accurate to state that this is the case with regards to the specific experimental form and moment of tuberculosis in people of the age group targeted, notwithstanding the U.S. context in which the trial was conducted. The representational powers of the RCT, however, often enable the 'discovered' treatment to be extrapolated to eventually allude to its efficacy in all biological bodies affected by the targeted disease. Streptomycin did not have such a straightforward path, and its side effects as well as the found development of bacterial resistance led to the adjustment of its use as a treatment, eventually being combined with para-aminosalycylic acid, which was shown to be more effective through three new trials (Daniels and Hill 1952). Following streptomycin and para-aminosalicylic acid in 1949, isoniazid (1952), pyrazinamide (1954), cycloserine (1955), ethambutol (1962) and rifampicin (1963) were introduced as antituberculosis agents and in different combinations, still with continuously emerging resistance of tuberculosis to these antibiotics.

Looking into *A. afra* in tuberculosis followed a similar path, with new hopes triggered in indigenous medicine, combined with a resurgence of tuberculosis associated with HIV/AIDS. The deaths attributed to tuberculosis versus those attributed to HIV/AIDS are difficult to untangle. In the Cape, the tendency is to diagnose the cause of death as tuberculosis, hence diminishing the importance of HIV/AIDS as the greatest killer. It is always difficult to attribute a single cause of death due to the multifarious facets of life in which it is embedded. 'Medical classification work as based on the International Classification of Disease (ICD) does not, however, give a context, it records a fact (one died of the disease or not)' (Bowker and Star 2000: 172). While this makes things simpler for the records, it does not attest to life's intricacies. Contexts play an important part in how a disease develops, is lived and resolved (or not), and how a bacterium takes its toll on a person's life. 'In life' the 'disease itself' is never isolated as in the laboratory, nor does it ever exist as a separate entity with the same results in all bodies.

Bowker and Star (2000: 171) mention that in the nineteenth century tuberculosis was qualified as a poetic illness, a disease of the 'sensitive', often thought an ideal, lady-like disease for middle-class women. The medical disbanding of these poetic aspects of tuberculosis from 'core' physiological mechanisms attests to hierarchies of knowledge that in turn define how a disease will be tended to, as well as lived. As is the case with knowing medicine, it is not enough to know disease in a controlled environment; the ways people live with medicine and disease are perhaps as telling, as are ways of engaging with them through skilled performances. In fact, when looked at up close, a disease loses its biological coherence, which solely appears and is made to appear in a controlled environment such as in the laboratory. The RCT demands such a closed notion of 'disease', usually only a specific experimental form that can be fixed in time and reproduced in the laboratory. The RCT model involves a notion of disease in the biological body, separated from its manifestations in multifarious contexts 'in the world'.

Medical anthropologists have found the need to expand the category of 'disease' to attend to its lived and felt experiences (Fassin 2000: 100). Fainzang (1989: 11), for instance, defined 'disease' as a triple reality made up of 'disease', 'sickness' and 'illness': disease constitutes a measurable, quantifiable entity or an organ or a system, while illness refers to how a person feels or to the lived experience of disease; the term sickness, as suggested by Laplantine (1992: 20–21), can refer to the 'process of socialization of disease and illness' (also mentioned in Roy 2002: 18). These newly added layers, however, leave the core entity of disease intact, even sustaining its credence or objectivity. Rather, it is other approaches that engage in the biology of disease in life that make its multiplicity appear. Mol (2002), for instance, does so by exploring what she names 'doing disease'. More precisely, she shows how arteriosclerosis is done in practice by a variety of actors implicated either as they live or attempt to heal disease in the hospital, actors only very seldom bringing arteriosclerosis into being as an isolated object in practice. While I did not follow how tuberculosis was being done, I follow 'doing

25

medicine' in a similar approach. I thus assume tuberculosis is also being done in multiple ways, while the first controlled trial to test a drug against tuberculosis shows how the process requires narrowing down the forms of tuberculosis as well as setting closure to the multiplicity of tuberculosis to conduct the experiment. Disease in this framework requires a universal experimental type. Only with this in hand and a molecule in view can the trial move forward. 'Indigenous knowledge' and its corresponding dignity of a people have much less leeway within this process; however, it fully emerges in practices, as the following chapters will show. I now, however, return to the ways this unravels with A. *afra* in the preclinical process: a strain of experimental M. *tuberculosis* being already fixed for testing, one needs to find such a constant within A. *afra*. A first toxicity study by Mukinda and Syce (2007) aimed to find a proper dose of A. *afra* in liquid extract to test for such a purpose, while a second study by Ntutela et al. (2009) adjusted the results of the 2007 study to instead test A. *afra* in animal feed against an experimental form of tuberculosis.

A. *Afra* on Trial in Tuberculosis

The link between A. *afra* and tuberculosis appeared in scientific publications in 2002 via investigations on the aqueous extract of A. *afra* indicating that the plant has bronchodilator and possibly anti-inflammatory activities (Harris 2002). 'Other assays carried out on the ethanolic and dichloromethane extracts have shown the plant to have *in vitro* hypotensive and anti-tuberculosis effects, respectively (MRC and SAHealthinfo 2004)' (Mukinda 2005: 2). Mukinda and Syce (2007: 138) state that such studies support the merits of traditional use. While they do so in an indirect fashion, none of these studies attempt to know *umhlonyane* as it partakes in everyday healing. All efforts are set in isolating and identifying possible chemical constituents such as, in this case, flavonoids. The safety study of A. *afra* in tuberculosis conducted by Mukinda and Syce (2007) is defined with interests in the investigation of the potential toxicity of the flavonoids contained in the plant. Before I enter into the details of the safety study, some broader aspects of the endeavour seem useful to delineate.

In studies done on the molecular configurations in plants, it appears that with every precision at the chemical level, as many new complexities seem to arise; the demonstrated mechanisms express potential clinical usage that may or may not be made useful for everyday use, and all imply dependence upon a biotechnological process only available through laboratory techniques and eventually an industry and market to make them accessible. The multiplicities or realities of A. *afra* in the world also reappear in different instances, such as when comes the moment to decide how to prepare A. *afra* to reach a precise dosage required to test the compound's toxicity level and to measure its safety, which can then be tested for its efficacy. A. *afra* is to be known in a precise manner and thus prepared correspondingly.

To justify an eventual trial, A. *afra* needs to be prepared into a constant dosage; thus, its benefits for its environment of growth must be controlled towards a single orientation as much as possible. The particularly high variability of A. *afra* was and remains a great challenge. Plants prepared for TICIPS' trial were initially grown on two different farms, but this had to be narrowed down to a single farm and farmer, since both the different microclimates and the particular way the two farmers nursed the plants created too much variability. This process shows the engagements of plants within mediums and the engagements of people with plants play into their very molecular compositions as well as how it is these connections and contexts that are required to be minimized for the purposes of the trial. In these practices, it is the human actions upon the plant that are to be reduced, as well as the different 'natures', which need to be decreased to only one. The selected farm and farmer is thus a compromise meant to facilitate homogeneity needed for the eventual RCT. In this compromise, A. *afra* is grown in ten rows, preferably all shoots from a single bush, again with the objective to reduce variability.

It is important to note here that a compromise was also reached in the reverse direction; while the RCT and ethical committees approving the procedure classically demand the isolation of a single molecule, the safety study upon which the preclinical trial will rest aims to test a whole plant and its molecular synergies[6], showing molecular biologists' interest in the 'chemistry of life', notwithstanding the legislative procedure of the RCT and further correlating with the work of Xhosa *izangoma,* as I will discuss in chapter 6. Agreeing upon how to tend to *umhlonyane* to ensure efficacy in healing, however, remains 'worlds' apart, foreshadowing some of the difficulties in dealing with indigenous medicine, which pull away from the already challenging task of isolating *A. afra* from the environment in which it found ways to survive and prosper.

A fundamental dissonance appears around the very *A. afra* bushes cultivated for TICIPS' trial. The farmer nursing the *umhlonyane* plants for the purposes of the trial knew very well of the common usage of the plant by the *izangoma* and had grown more than required for the trial, assuming he could also sell the plants. The *izangoma,* however, refused the *umhlonyane* growing on the farm chosen for the trial, stating it had lost its 'life' and thus its efficacy. The life and efficacy of plants for healers is created through establishing good relations with the plants; namely, by letting the plants provide their benefits within their contexts of growth rather than through controlled measures or abnormal environments to which the plants are not adjusted. Cultivation techniques, plantations and farmlands are not neutral grounds even in the postapartheid context, adding to a fundamental disagreement with an externalized positioning towards plants.

Proximity with the plant is mentioned as a source of knowledge legitimacy by the healers, whether the plant be grown in the yard, collected in the 'wild', dried and kept in the house or known to grow in this or that area. It is not solely physical proximity, such as being born and bred with plants and living with them, as a Xhosa *isangoma* explained; knowing healing with plants is also received through a 'calling'. The most common calling of becoming an *isangoma* across much of southern Africa is 'the sickness calling' (*ukubiswa* in Xhosa) (Wreford 2008: 104), usually a condition of both physical and emotional distress that inhibits one from continuing on with life as usual, plunging the initiate into explorations to learn, act and eventually heal with ancestors. I will delve into the details of this highly ritualized initiation done with an *isangoma* mentor over the course of decades in chapter 6; it is enough here to simply signal that it is a process of deepened engagements in both past and present worlds that enable *izangoma* to activate plants at a distance, learning to heal with and through them. It is lived and felt proximity with plants that presides in *isangoma* practices and that corresponds with South African Rastafarian *bossiedoktor* practices as well, albeit done in different ways.

Rastafarians *bossiedoktors* in the Cape, to whose practices I return in chapter 5, explain knowing plants as given by Jah (God the

Creator), as is the case throughout the movement that emerged in Jamaica in the 1930s; however, in South Africa knowing plants also comes from elders, in this case the KhoiSan[7], also known as San Bushmen from the Kalahari or simply the San (Laplante 2009b, 2012). The San people Gibson met during her study in north-eastern Namibia 'stressed that plants breathe, live, reproduce, feed, poison, defend against or hide themselves from predators. They "drink", "eat" and move in the air. They can travel across distance... Plants can be stronger or weaker, depending on where they grow. Through the movement and sharing of *máq,* i.e. essence/wind/ air, people, animals and plants interpenetrate and transform each other' (2010 : 57). This open-endedness given to plants are also echoed in Xhosa *izangoma*'s ancestral practices as will become clear in chapter 6.

The common appeal to indigenous African roots explains in part how *izangoma* and Rastafarian *bossiedoktors* agree upon similar ways of tending to plants. The two healer groups for instance generally agree that the ways of manipulating plants in the preclinical pro-cedures, namely, through cultivation, make the plants lose their ef-ficacy. Contrary to the practices of the preclinical trial, *izangoma* engage with plants as a means to heal rather than assume plants contain parts that heal in and of themselves, independently from the ways we engage with them. Like the variability valued in the plant, the healers also cherish proximity and entanglements with the particular therapeutic problem. As such, *izangoma* often rely upon Rastafarian *bossiedoktors* to obtain medicinal plants, because they share with them an understanding of the world as a sentient being. It is through life that plants work and in life that they are to be found beneficial in healing. In both cases it is through deepened en-gagements with the world and things that healers resolve problems, posing numerous challenges to the RCT standard.

The explicit intention to withdraw from the world to 'know it' through the experiment is thus the source of profound dissonance within the preclinical trial. The little I have described of the steps leading up to the preclinical process already shows how nature is not something 'out there' from which to extrapolate molecules, yet it is something that is made to be understood this way in particular prac-tices. These instances further show how we are always dealing with multiple natures and cultures in practice, as acknowledged through the process of reducing interest in *umhlonyane* to that cultivated on a single farm and by a single farmer. As is done with the placebo in the clinical process, rather than attending to engagements of people with plants in modifying their constituents and thus their efficacies, these engagements are to be excluded from what is to be known about *umhlonyane*. The placebo becomes the emblem for all the healing occurring in the disguised "no treatment" arm of an RCT: '[a]nything that threatened the fastidious detection of a predictable cause and effect outcome was conveniently disposed of in a repository labelled the "placebo effect"' (Kaptchuk 1998b: 1723). Paradoxically, it is precisely the acknowledgement of the 'power' of the placebo to have an effect on healing and of the 'power' of the farmer or healer to have an effect on the plant that leads to their marginalization.

In privileging 'objective' knowledge, the placebo has slowly become equated with 'bias' or a threat to true scientific evaluation. The newer meaning of placebo as 'bias' is one crafted within the RCT experiment. With this displacement of placebo to the sidelines, specifically because of the acknowledgement that it is active and 'powerful', 'the dummy side of a trial received inadequate attention, poor methodology, and a priori assumptions' (ibid.: 1724). The obsession to eliminate placebo (equated to 'bias') in the experiment had the effect of reducing the RCT to a process of the evaluation of 'natural' effects, essentially physiological and molecular ones; in this move, the quest for objectivity and an objective method came to preside over the quest for a therapeutic outcome. All that resists objective knowledge of a certain kind, even if it shows efficacy, has thus been largely disregarded for the past sixty years, only to reappear recently as something worthy of investigating. Lévi-Strauss (1949) had earlier attempted to understand 'symbolic efficacy', a line of research that anthropologists have continued to pursue, showing, for instance, that 'the placebo effect is the same as the effectiveness of bodily-experienced symbols' (Ostenfeld-Rosenthal 2012). Attending to these intricacies implies 'knowing from the inside', while even the steps leading up to the preclinical trial are bound to 'knowing from the outside'. The intention to disengage from the plant to know it is a locus of dissonance stemming from the demands of the RCT model downstream rather than from the scientists working within the trials who pull towards the ways the plant is used in context as much as possible.

Mukinda and Syce's study of A. *afra,* is for instance concerned with the the safety of the plant which is not reported in the botanical inventories, yet they prepared an aqueous extract of A. *afra* 'in a way that, as much as possible, mimicked the method traditional herbal practitioners use to extract their plant medicines' (2007: 139). As shown above, the ways that A. *afra* is reported to be used, and what I have witnessed myself while attending to the plant, vary enormously. For example, in my own observations with a family of Rastafarian *bossiedoktors, umhlonyane* was taken directly from the bush in the backyard and rubbed with oil on the stomach of a feverish baby. I was also given part of the foliage to taste fresh, stating it tasted like 'medicine'. I was invited to smell the plant, known for its aromatic virtues. During another visit in their home, a tea infusion was pre-pared for us, mostly for taste. Stems of the bush were also dried and bundled for sale in the markets. A Rastafarian *bossiedoktor* whose herbal stall I visited numerous times always had a bundle of *um-hlonyane* for sale, explaining people who bought the plant already knew what do to with it and, if not, he offered guidance in ways of preparing it. An *isangoma* explained that *umhlonyane* stems were placed under the bedsheets of the patient the day before a healing ritual for purification purposes, namely, to clarify dreams to be more open for healing. Another *isangoma* explained *umhlonyane* makes dreams clear and it can also be used alone for the stomach or with *imphepho (Helichrysum odoratissmum* or *H. petiolare,* also referred to as a dream herb) or pepper for cough, cautioning however that it does not do this by itself yet depends on the ways one connects with the plant. These are some of the specific engagements with *umhlo-nyane* to which I will return to in-depth in the following chapters. When I spoke to Mukinda and Syce, they explained that they had mimicked the use of A. *afra* in the form of a tea, or an infusion, which they concluded was the most common usage, notwithstand-ing the diversity of preparations.

The botanical inventories I have described above provide infor-mation on the uses of A. *afra* extrapolated from their contexts. It is within this enumerated multiplicity of tapped and codified uses in the everyday that a choice was made as to how to test its safety in tuberculosis in the form of tea. These common uses are documented and hence 'known'; however, they are not 'known' in a precise way recognizable through the current scientific route, which needs to be done through analysis of the chemical constituents. It is only then that we assume to 'really know', although that knowledge will move further away from the everyday until it is reintroduced in a new form should *A. afra* become a biopharmaceutical, for instance. What is to be gained with the acquisition of precise demonstration of the real efficacy is at the cost of disengaging this knowledge from its contexts and, eventually, potentially making it unavailable for everyday usage: doubt may be instilled for fear of its potential 'unknown' toxicity, which is relegated to a higher legislative order as opposed to being known through proximity.

A main result of the study of Mukinda and Syce (2007) was in fact to discredit the practice of the usefulness of *A. afra* in tea form, since it was found not to work on tuberculosis in liquid form. This can discredit either 'historical use' or the researcher's decision to use *A. afra* in tea form. When I asked the first author how they had decided to test the plant in tea form, he pointed out information found on *A. afra* on page 142 in van Wyk and Gericke's book *People's Plants* which mentions 'fresh and dry leaves and young stems are usually used as infusions decoctions and tinctures' (2000: 142). A second such inventory was further mentioned, one with a detailed receipe, also extrapolated from its context: 'The most popular traditional method for preparing a tea (infusion) or decoction is to add a quarter cup of fresh leaves to one cup of boiling water, allow it to stand and steep for 5 minutes, then strain and sweeten the filtrate with honey or sugar to mask the bitter taste. Thereafter, the tea is sipped at a set time (to ease all types of cold, cough, heartburn and bronchial disorders) or used as a gargle (e.g. for a sore throat)' (Roberts 1990: 227). The second author further mentioned he had enquired generally about the use of *A. afra* in his surroundings and concluded it was most often used in the form of tea. In line with the current laboratory practices, Mukinda and Syce (2007) for their part applied an early use of placebo tea bags to test *A. afra*, to eliminate the 'bias' that may arise in the simple act of drinking tea as therapeutic, notwithstanding the bioactive compounds. The placebo control as 'bias elimination' is the precise procedure that dismisses acquired skills to make medicine work, in this case the act of 'drinking tea', which can be an art form and therapeutic social process, as is currently the case in Japan and elsewhere. While traditional use is considered and orients some of the preclinical tests, it is to dismiss it rather than to give it legitimacy, and even has the potential to render its use illegal. The placebo is otherwise not normally brought into being in its strict sense during the preclinical trial. Numerous other aspects of the world are implicitly discarded as 'placebo effect' or 'bias' in a broader sense.

To meet the standards of the envisioned RCT, the plant was obtained from a specific location; freshly picked A. *afra* plant material was obtained from Montague Garden (Western Cape, South Africa) during its flowering period in the summer of 2004 (Mukinda and Syce 2007: 139). It was afterwards authenticated by the Department of Botany of the University of the Western Cape. The specific bioactivity selected needs to be isolated from the plant as well. To maximize the activity of the selected compound, the plant is prepared in a certain way. The wet leaves of A. *afra* plucked from the stalks were washed with distilled water and dried in the oven at 30 degrees Celcius; the dried leaves were suspended in distilled water (fifty grams of dried leaves per one litre of water) and the mixture boiled for thirty minutes. The decoction obtained was cooled, filtered, frozen at −70 degrees Celsius, freeze-dried (using a VirtisTM mobile freeze-dryer, model 125L) and then sterilized by gamma irradiation. 'The freeze-dried aqueous extract of *Artemisia afra* provides a form that can be standardized in terms of constituents (marker compounds) and physicochemical properties and is easily packaged, stored and/or converted into high quality pharmaceutical dosage forms' (ibid.: 142). It is, however, not accessible otherwise.

Healthy BALB/C mice were simultaneously kept in a controlled homogenous environment, supposing this would be more telling. Acute toxicity studies were done in these mice, the most important parameter being the measure of death after a specific dose. Chronic toxicity studies were done with male Wistar rats of a particular weight, supposing seven days is sufficient for them to acclimatize to laboratory conditions (and not create bias due to adjustments to a new environment). Weights were recorded and samples of blood were collected from the individuals. Collectively, the results of the Mukinda and Syce (2007) study indicate that the aqueous extract of A. *afra* is nontoxic when given acutely, has low chronic toxicity potential and, in high doses, may have a hepatoprotective effect, notwithstanding the multiplicity of ways living organisms might engage (or not) with given stimuli in an abnormal and controlled environment. Overall this study paves the way for the planning of future preclinical and clinical studies of this plant medicine, namely it opened the possibility for the preclinical study of A. *afra* in tuberculosis to take shape.

Preliminary Verdicts

The preliminary results of the preclinical trial study about A. *afra* appeared in 2009.[8] The study by Ntutela et al. (2009), implicating Mukinda and Syce respectively as third and thirtieth author as well as five other leaders of the TICIPS study, looked into the efficacy of A. *afra* phytotherapy in experimental tuberculosis. I met more than half of the fourteen coauthors of this research in both South Africa and the United States during my research and I delve into more details of their practices beginning in chapter 3. I here want to briefly mention what is made to appear as legitimate knowledge within such publications, how they build upon previous inventories and studies as well as how they improvise new ways of knowing the plant rather than simply 'discover' or 'uncover' its mechanisms. Not only are such publications embedded in multiple locations of herbal knowledge, yet there are many scientific contributors partaking to the quest of making a biopharmaceutical, notwithstanding technicians and other's whose names will never appear (Osseo-Asare 2014). The 2009 study by Ntutela et al. led to an unexpected finding that once again is stated to support the merits of traditional use of A. *afra*, this time including treatment for respiratory ailments.

Ntutela et al.'s study 'investigated the inhibitory potential of A. *afra* extracts against M. *tuberculosis*' (2009: S34). The 2009 study reports on antimycobacterial properties of A. *afra* on M. *aurum* and M. *tuberculosis*, showing enhancement of immunity against tuberculosis when particular active fractions of A. *afra* are given in animal feed. The aqueous extract tested for toxicity in Mukinda and Syce's 2007 study had not shown this activity; it had shown a regulation of pulmonary inflammation during early infection. This is explained by how the extract of the plant was prepared (boiling it for thirty minutes, amongst other procedures), which may have annihilated the antimycobacterial effects. The decision to test the plant in the form of tea was explained as resting upon the ways people mostly use the plant. The findings of the Ntutela et al. research however 'would suggest that such active ingredients would not be extracted in traditional practices where application is usually associated with boiling of plant material in water' (Ntutela et al. 2009: S38). This could paradoxically discredit the traditional way of making A. *afra* as a tea since the plant would have water-insoluble active components, while recognizing its common use for respiratory ailments when combined with food. At least it has shown such activity when administered as an integral component of a formulated feed through this specific experiment.

In the 2009 study, the researchers found that a dichloromethane extract of A. *afra* does contain antimycobacterial activity that can inhibit both rapid-growing M. *aurum* and virulent M. *tuberculosis* replication when prepared in animal feed. Dried leaves (two hundred grams) were suspended in distilled water (four litres) and the mixture was boiled for thirty minutes. The preparation obtained was left to cool before it was filtered and the filtrate freeze-dried. The plant material was then extracted in dichloromethane, an organic compound used as a solvent. The mixture was stirred for three hours and left to stand overnight. The extract was recovered by filtration; the solvent was removed using rotary evaporation. The plant material was reextracted two more times and all three extracts then combined together. The extracts were then fractionated to isolate the active compounds; I skip a good number of other technical procedures yet the key to the success of this experiment seems to lie in administering the phytotherapy to animals as feed. The sophisticated technical procedures and precise documented steps are useful to pursue with further experiments. These procedures also attest to entirely new (dis)engagements with the plant, done via precise repetitive acts involving machinery. While these practices are assumed as 'standard' and 'objective' from a scientific point of view, these procedures and those done in the 2007 study are part of a larger set of scientific laboratory practices that makes them even thinkable.

In the 2009 study, white albino mice (C57B1/6) were prepared for the experiment. They were bred and maintained under specific pathogen-free conditions. Female mice, between the ages of eight to twelve weeks, were used in the experiments. In a separate laboratory, M. *tuberculosis* H37Rv was grown. The issue of the different types of tuberculosis was resolved by using an experimental form of M. *tuberculosis*, H37Rv, which was injected into the mice. Once infected, the mice were given the A. *afra* infused animal feed. They are then 'sacrificed' at the indicated time points. 'Sacrifice' is the term used by scientists to refer to the moment of killing the mice. 'The term "sacrifice" is used by experimental biologists to describe methods for killing laboratory specimens' (Lynch 1988: 265); it is 'a name for methods of killing an animal so as to transform it into data' (ibid.: 274). Organs are removed and homogenized for further study. This way of proceeding with living beings seems to dissociate types of life by degrees of 'nature'. 'Sacrifice' is also a term used within Xhosa *isangoma* practices, explained in detail by Wreford (2008) as a way to complete initiation; a goat and a bull are sacrificed, but they are given proper burial, ensuring connection with the ancestors involved. The C57Bl/6 mice, however, are treated as objects.

The particular relation or lack of relation with other living beings that are part of the trial is permitted through a strong sense of 'science' for truth, for the well-being of 'humanity'. The C57Bl/6 white albino mice are kept in a cage in the back of the laboratory, treated 'equally', assuming they should all react in the same way both to the way they are (mis)treated and to whatever will be injected into them. It is assumed that it is possible to observe, in a shared environment, how the mice's bodies react first to the disease injected into them and, afterwards, how a specific dose and preparation of *A. afra* acts upon the disease, reducing its effects (or not) in the lungs and tissues. Researchers can in this manner, upon sacrificing the mice and visualizing the effects of the disease and of the plant extract, account for precise understanding of some effects of *A. afra* in these particular conditions and on particular parts of bodies. In this case, *A. afra* was found to modulate pulmonary inflammation (Ntutela et al. 2009). Bioactivity in *A. afra* is studied in mice to later understand how it instructs bodily molecular configurations in humans.

Birke (2012) makes important parallels comparing the objectification of animals to that of human bodies during clinical trials. The body, whether of the animal specimen, the human or, in this case, the plant, are all turned into analytical objects for the time of the experiment, standing in stark contrast to the living, experiencing body that we will see permeates the heart of indigenous medicine. Of concern is how these procedures continue when they are performed on human subjects, although with less control and without the privilege of sacrifice. In either case, all the living beings involved (plant, animal, human) in the experiment are somewhat temporarily 'deadened' in the sense that they are given no abilities or skills to engage with newly introduced information; rather, they are assumed as passive until reception of stimuli. It is through these procedures that the researchers concluded that *A. afra* is a viable source for identifying antimicrobial compounds and contains anti-inflammatory substances that may potentially be useful for clinical application. They propose a new application model where water-insoluble phytochemicals can be applied successfully in therapy (Ntutela et al. 2009: S40). In this way new possibilities are opened for potential future clinical applications.

A few aspects of the preclinical processes mentioned above are of interest for my upcoming discussions. First, what is learned about the safety of *A. afra* is restricted to a sole compound as well as a specific method of extraction and preparation of the compound. Second, although historical use of *A. afra* is given, pursuit of the preparation of the plant is rather oriented towards its defined goal of finding a cure for tuberculosis; testing it in the most common tea form subsequently discredits this use, at least with regard to its benefits for inhibiting an experimental form of tuberculosis. Third, all steps of the preclinical trial turn its participants (plant, animal, human) into an 'object-in-general' (Merleau-Ponty 1945: 1)[9], an analytical object disbanded from the lived body. Finally, what is known in the end are mechanisms and new possibilities; one such new possibility is an unexpected anti-inflammatory activity of the aqueous extract, and the second is a new model for testing hydrophobic phytotherapies in animal feed. This flexibility relies upon awareness of multiple potentials of the same plant depending on the ways it is prepared. The promise of closure of the experiment appears to be overridden with the opening to new possibilities. While the RCT, or its preparation, can be explained as a way to get something done, or to move forwards in the process of making a medicine, it is shown to be extremely partial, situated and specific, although it claims in this way to represent a universal mechanism of a bioactive compound in human bodies and eventually as a solution to a world pandemic. Once complexities of 'real life' are reduced so as to enter the complexities of 'laboratory life', the plant has been greatly transformed; in its specificity and controlled dose, it can, if the results are positive, be put through the RCT process and potentially reach this global objective.

While it can be argued that the procedure as such 'works' and produces 'effective medicines', I argue it is in no way a straightforward path. First, the mechanism can be shown to work in the laboratory as well as on large populations, yet often only as 'the next best thing'; in other words, showing greater potential than the current protocol. Further, if we are reminded of the specificity of the experiment, this 'effective medicine' supposes all biological bodies are the same, notwithstanding the contexts in which these people live. The latter will, however, play into the potential efficacies of the medicine. Finally, if medicine is to be effective 'in itself', the disease and bodies it is meant to respectively counter and heal should be fixed objects, however they are continuously emerging in new ways.

On Making an Object

> Only we who have erected the objectivity of a world of our own from what nature gives us, who have built it into the environment of nature so that we are protected from her, can look upon nature as something 'objective'. Without a world between men and nature, there is eternal movement, but no objectivity.
>
> – Hannah Arendt, *The Human Condition*

The preclinical study of *umhlonyane,* in view of conducting the scientific gold standard, an RCT, relied upon an accumulation of corresponding forms of standardization of bodies and of disease. Scientific literature has for instance conveniently made A. *afra* into a botanical plant entity classified in a kingdom, sub-kingdom, division, class order, family and species etc. In this life form, A. *afra* can thus be distinghished from other similarly classified life forms as well as it can transcend space and time. It is A. *afra* whether it is in a township backyard, powdered in animal feed, placed under bedsheets, drunk in tea, snuffed, tasted, eaten by an infant or a monkey. I have shown a similar process occurs with M. *tuberculosis.* These standards, classifications or categories are nested inside one another as they act in concert, manipulating things and living beings in a way that gives up living in them, as Merleau-Ponty found all of science to do (1945). It is a process of moving away from life so as to find medicine that will be useful in life. It does so by making living beings and things into

38

natural objects of enquiry. RCTs are currently believed to be the best standard available to ensure the safety and efficacy of medicines by a good number of people across research consortiums, laboratories, nations and technical systems. It tells of the kind of research that is found to be the most useful on the transnational world scene. The RCT is a deductive research model, one that is perhaps the most regulated by codes of ethics guiding research, and in this way is prescriptive of ethics and values that have great consequences for individuals.

The RCT standards have consequences for scientists who embody the RCT's practices of disengagement as well as for the human and nonhuman actors who temporarily become analytical objects of enquiry. As a legislative method in powerful world circuits, RCTs play a reassuring role for those who endorse its process and outcomes (even if blindly). However, as 'with all deductive methods, the benefit that the conclusions follow deductively in the ideal case comes with a great cost: narrowness of scope. This is an instance of the familiar trade-off between internal and external validity. ... There is no gold standard' (Cartwright 2007: abstract). In the still-favoured deductive approaches, the key is in the predesigned model. The search for data comes afterwards, sometimes to test the model. From their original objective of minimizing health hazards by testing therapeutic processes precisely, RCTs have narrowed their focus to proof of method and more recently to the insurance of the safety of a product. From another perspective, RCTs have become more than demonstration of method; they have become a way of life, a way to manipulate life, to decompose and recompose it in varying ways.

RCTs have, however, not become a true (re)presentation or discovery of 'reality'. What escapes the RCT procedure is precisely what it dismisses as placebo; hence, it misses ways of making medicines work meaningfully in life-making processes entangled within physiological and micromolecular pathways. The traces of the clinical trial that might be differentiated from trial and error 'in life' can be recognized through techniques such as 'control groups', 'random' methods, 'placebo control' and 'blind assessment', all aiming to 'discover true efficacy' of a medicine however doing so by disengaging with real meaningful engagements in life with medicine. From 1800 onwards, clinical trials began to proliferate and more attention was paid to study design. Placebos were first used in 1863, and the idea of randomization was introduced in 1923. In 1938, the word 'placebo' was first applied in reference to the treatment given to concurrent controls in a trial (Diehl et al. 1938: 1168). Clinical trials blossomed after the Second World War as a way of strengthening the science of drug testing by attempting to eliminate social responses to treatment (Löwy 2000: 50; see also Saethre and Stadler 2010: 99). In Britain, the RCT became one of the tokens of the social revolution that changed British medicine forever and for the better at that time as acknowledged by Sir John Grimley Evans (2010: 239). The National Health Service, created in 1948, was to lead the war on disease within a particular social context in which it made sense.

A similar phenomenon occurred in the United States during that period that redefined the shape of medical research. The 'golden years' of the National Institutes of Health (NIH) expansion are between 1955 and 1968, as the budget expanded from $8 million in 1947 to more than $1 billion in 1966 (NIH 2012b). Thirty years later, in 1998, the National Center for Complementary and Alternative Medicine (NCCAM) was established by Congress, with the objective to 'investigate and evaluate promising unconventional medical practices' (NIH 2012a). It adopted the RCT method diligently. The RCT has since become the gold standard for the evaluation of pharmaceuticals, biologics, devices, procedures and diagnostic tests (Friedman et al. 1998), as well as, more recently, 'indigenous medicine'.

In the process, the usage of RCTs has become varied; however, the core features, even if reassessed and refined, remain embedded in the model. There are explanatory trials asking whether an intervention works (or not). They are often combined with elements of pragmatic trials testing the broader consequences of an intervention in daily life. There are efficacy and effectiveness trials. '*Efficacy* can be defined as performance of a treatment under ideal and controlled circumstances' (Revicki and Frank 1999: 423). '*Effectiveness* refers to the performance of a treatment under usual or 'real world' circumstances' (ibid.: 424). To the basic science trial (how does it work), the efficacy studies and their outcomes (what are the effects) and effectiveness research (how well does it work in real-world settings), translational research has recently been added, asking if it can be studied in people. Translational research is about identifying and validating biomarkers or other signatures of biological effect; it is about developing and validating measures of outcome, validating treatment algorithms and measures of quality control. It also concerns developing preliminary clinical evidence regarding efficacy and safety, and establishing feasibility or estimates of sample size for future studies (NCCAM 2011: 11). The emphasis of clinical trials, however, largely remains on basic research and efficacy studies.

It is currently the double-blind RCT that holds the status of the 'proper' RCT. The RCT 'seeks to confer the ideal of scientific exactitude onto clinical experimentation in an effort to attain the objectivity of the laboratory model. A double-blind RCT is considered medicine's most reliable method for "representing things as they really are" ' (Kaptchuk 2001: 541). Rorty (1979) who is paraphrased in the previous quote, highly critiqued this confusion in the way 'objectivity' was understood to mean both rational agreement and the mirror of nature (1979: 334). The RCT's claim to simply mirror nature however rests more upon a rational agreement in a pre-designed model enabling to control life than it does upon an understanding of ways to move to the rythms of 'nature' beneficially (and thus mirroring it). The preclinical trial through which I travel in this book was in preparation for an efficacy trial along the lines of a 'proper' RCT; it was asking the research question of how it works, to only later deal with effectiveness in real-world settings in order to confront some of the complexities arising from the introduction of new technologies into lived contexts. The preclinical trial already greatly strays from the lived experience of therapeutic processes, even it is its main task. In this way it disregards most of what is done in indigenous medicine, as well as dismisses new approaches within the sciences that it legitimates.

With recent developments in genetics and molecular biology, the borders between nature and culture have greatly shifted, giving new meanings to bodies, disease and other forms of life. According to Tylor, scientific thought as maintained through an RCT was already an archaic mode of consciousness surviving in degraded form in the 1980s since it had outdated its own moment of emergence (1986: 123). It is always a question in meetings with varying groups of experts on clinical trials whether more forms of standardization need to be added to palliate yet another 'bias', or whether the whole experiment needs to be reassessed for what it actually provides as information. In other words, does the RCT demand more legislative methods or more flexibility? I aim to grasp how such procedures are done in practice, how other ways of making medicine work are done in indigenous medicine, as well as eventually explore what it could look like to do medicine otherwise, without making objects lifeless in order to understand treatment, therapies and efficacies in the movements in life. The preparation for the testing of an indigenous medicine through an RCT brings these issues to the forefront, making it an ideal subject of anthropological enquiry.

Notes

1. Febrile complaints, or *umkhuhlane* in Zulu, are described as one of the most common ways to refer to 'any general constitutional derangement of a febrile and generally infectious nature, and may include enteric, scarlet and malarial fevers; small-pox and measles: pneumonia, acute bronchitis and influenza, as well as all the commoner minor catarrhs and bad coughs to which one is periodically liable' (Bryant [1909] 1966: 51–52).
2. Following experimental administration to rabbits, the volatile oil of *A. afra* has been isolated by Van der Lingen and has the reputation of being as toxic as oil of Sabine (Juniper oil), producing hemorrhagic nephritis, degenerative changes in the liver and pulmonary oedema (Watt and Breyer-Brandwijk 1932). The toxic and hallucinogenic effects of this volatile oil are attributed to thujone, so that overdose or continued use over long periods of time is potentially harmful (van Wyk et al. 2000). Volatile secondary metabolites are found to vary frequently due to geographical variation, with regard to the different plant parts used, drying methods and variation within the natural population (N. Liu et al. 2009).
3. Flavonoids are reported to have many potential activities, such as antioxidant, anti-inflammatory and antiallergic (Harborne and Williams 2000). Additionally, they have chemopreventive activity against skin cancer (e.g., apigenin); inhibitory effects on chemically induced mammary gland, urinary bladder and colon carcinogenesis in laboratory animals (e.g., hesperetin); and anticarcinogenic and platelet antiaggregatory effects (e.g., quercetin) (Erlund 2002; Waithaka 2004). Furthermore, the flavonoid luteolin has been shown to exhibit antimutagenic and antitumorigenic activities (Shimoi et al. 1998), as well as vasodilatory and potent antiplatelet activities (Harborne and Williams 2000). It is also a promising agent for use in ophthalmology for the prevention and treatment of cataract and vascular eye disorders (China Great vista chemicals 2002). According to Mukinda (2005: 3), no adverse effects have so far been reported that might specifically implicate the flavonoids of *A. afra*.
4. The first published RCT appeared under the authorship of the Streptomycin in Tuberculosis Trials Committee in 1948. One of the researchers of this committee, Sir Austin Bradford Hill, is usually given credit for the modern RCT (Stolberg et al. 2004: 1539). Hill's pioneering achievement is to have brought forth the methodology of the RCT and applied it in

subsequent studies following the Tuberculosis trial (1952, Daniels and Hill 1952).

5. 'Since 1945, the ethical impact of clinical trials has become increasingly important, resulting in strict regulation of medical experiments on human subjects. These regulations have been enshrined in documents such as the Nuremburg Codex (1947) and the Declaration of Helsinki (1964, amended in 1975, 1983, 1989, 1996, 2000 and 2001)' (Twyman 2004). RCTs have thus become about safety and efficacy, aiming to insure a balance between medical progress and patient 'safety'.

6. The synergy between the compounds of the medicine as a whole was also taken into consideration in the clinical trials of the indigenous medicine *Sutherlandia frutescens* conducted in Kwazulu-Natal, South Africa (Gibson 2011: 134).

7. The KhoiSan people of the Northern Cape are descended from San hunter-gatherers and the later-arriving KhoiKhoi. Their language is distinct from the other African Bantu languages. KhoiSan is also a general term which linguists use for the click language of southern Africa. See http://www.khoisan.org (accessed 28 August 2014).

8. A separate preclinical study is looking into *A. afra*'s antimalarial properties, since such properties have been found in the related traditional medicinal plant *A. annua* L. (family Asteracea) known to contain artemisinin (Avula et al. 2009). Artemisinin has been shown, through RCTs, to be active against *Plasmodium falciparum*, especially against chloroquine-resistant strains. Artemisinin is currently an important compound of a novel class of antimalarial drugs used as a protocol medicine in global health networks and 'considered as part of the ideal strategy for malaria in Africa by WHO' (C.-Z. Liu et al. 2007). It has not been found in *A. afra*.

9. All translations from foreign material are my own.

Chapter 2
Engaging in Medicine

Science manipulates things and gives up living in them.
– Maurice Merleau-Ponty, *La phénoménologie de la perception*

When I asked a Xhosa *isangoma* (healer-diviner) 'How does *umhlon-yane* work?', I was invited to be immersed in the world through drum sounds. It shows how a simple question can be answered in very different ways. I had designed my start-up question to get at the heart of the preclinical process I aimed to understand. It is only during a conversation with the lead pharmacologist involved in TICIPS' pre-clinical trial of *Artemisia afra* that I learned he aimed to answer this question as well. He explained exactly this: 'How does it work' is the basic science trial question. The indigenous healer's answer however led me to add 'the world' to my question, which became 'How does *umhlonyane* work in the world?' This question will also be part of the clinical process, however, very much down the line; the complete randomized clinical trial (RCT) process may take up to seventeen years.[1] It is only in the final phases, once the new molecule or medicine has been made and tested for its effects in laboratory settings, that effectiveness research will ask how well it works in real-world settings. The beauty of the preclinical trial for me as an anthropologist is precisely that the object 'medicine' is not already done, and, in this case, it wants to be done by delving into indigenous medicine; or does it? While the preclinical moment is indeed a movement of opening into life-making processes, its intention is to withdraw the plant from the world rather than grow into it. The real question to ask both scientists and healers is therefore 'How is *umhlonyane* being done and undone to work in healing?', thus enabling us to attend to both the process of closure of the plant into the laboratory and to the process of opening of the plant through deepened engagements in-the-world.

With the hope of teasing out commensurabilities between the different actors' practices in the preclinical trial of *A. afra* and beyond its scope, I accepted the invitation to immerse myself more deeply in the world and learn from the plant. While I could have proposed a way to assess indigenous medicine to make them more recognizable in the preclinical procedure, for instance, by adding information about its technical use, collecting practices, recipes and preparation methods, this path did not seem fruitful, since it has already been undertaken by a good number of scientists, with botanical works filling up with lists of uses and constituents of *A. afra,* as I showed in the previous chapter. I hence wish to explore another route, one that indulges in the open-ended moment of the preclinical trial I had imagined, and one in which numerous scientists involved in the preclinical trial evoked as something they would like to do as well to really 'know' the plant in its environments. I thus propose a way to attend to the preobjective in both indigenous and scientific practices permeating the preclinical moment.

This chapter aims to delineate this approach. It takes root in medical anthropology, more precisely in its phenomenological turn towards 'embodiment'. Phenomenology also permeates the emerging field of the anthropology of the senses from which I took some inspiration, a theme that I will develop under the concept of 'sentience'. Finally, the importance of experiencing place/context in anthropology is also phenomenologically informed, converging with the reflexive turn in anthropology in the 1980s. This I discuss in the context of 'mediums'. A final section opens into a notion of 'worlds of becoming' since tending towards the living body takes us from medium to world. Before I entertain these enmeshed themes, I introduce my phenomenological approach in anthropology more generally with regard to its relevance to understanding the preclinical trial without giving up living in the things it manipulates.

A Phenomenological Approach in Anthropology

To inhabit the preclinical trial, by which I mean to 'know it' as experience, I take inspiration from a few foundational and enmeshed phenomenological insights in anthropology. I have mentioned already in the previous chapter how I adhere to Merleau-Ponty's (1945, 1964) phenomenology of perception, a key turn to broaden my understanding of what is going on in the preclinical trial of *A. afra*. The previous chapter thus precludes my approach with my positioning on 'knowing', which breaks from the naturalist or positivist ontology and thus refrains from foreshadowing objects made out as 'natural' or external as orienting the preclinical trial. This is also a core feature in the phenomenological project. One of the fundamental insights coming from phenomenology, which fits with anthropology, is that it 'necessitates the abandonment of the natural attitude linked to our experience and our thought, in other words, a radical change in our attitude' (Husserl 1950: 6). This 'natural' attitude can be understood as the need to think outside the ways science is currently organized, namely, as the experiment is designed through a separation of 'nature' from 'culture' or what Descola (2005) names a 'naturalist ontology'. This broad generalization can, however, be grounded in my own research trajectory.

When I began this research in 2006, it was with epistemological concerns with 'efficacy' linked with issues of knowledge legitimacies. To move beyond this discussion, which ultimately leads to the finding of incommensurability between science and indigenous knowledge, my epistemological concern widened to include ontological issues, and phenomenology provided the way to access them. I had fallen upon the term 'phenomenology' in the works of Kleinman (1980, 1988), Good (1994) and Hahn (1995) during earlier research with reference to their preoccupation with the lived experience of suffering and pain, yet I had paid little attention to it at the time (in 1998), perhaps unable to deal with these complexities. What convinced me to really engage with a phenomenological approach this time around was a combination of the very effective words of the *isangoma* mentioned above as well as a desire to understand more profoundly how this kind of knowing was being done. This implied taking on a new approach altogether. I simultaneously fell upon phenomenology in a turn towards works in the anthropology of the senses, which are similarly left unsatisfied with 'the restrictive understanding of the phenomenal world that is possible using the conventional descriptive instruments of an academic discipline' (Herzfeld 2007: 432). It is possible that a 'return of the body' as both a cultural and an academic moment (Vigarello 2007) is part of this resurgence of phenomenological approaches, however difficult they are to apply in academia. In its emphasis on immediacy, phenomenology is, however, likely a source for new creativity in anthropology and beyond, and it is also useful in this way.

Hallowell (1955) was the first in anthropology to present his work under the banner of phenomenology, followed only later by Csordas (1996) and Jackson (1996). Phenomenological approaches have recently been assessed by Katz and Csordas (2003) in anthropology and in sociology, followed by a more specific discussion of these approaches around the themes of experience and religion by Knibbe and Versteeg (2008) and more generally in anthropology by Desjarlais and Throop (2011). Ingold (2000, 2011), who is one of the strongest inspirations throughout my whole work, also borrows from Merleau-Ponty's phenomenology of perception in his anthropological approach.

First, a phenomenological stance does not begin with a priori 'objects', 'models', 'concepts' or 'protocols'; rather, it is the experience in the context that tells how these 'objects' are brought into being (or not). The objects are bracketed and experience is received in terms of what it feels like to move through the spaces where they appear. This is opposed to assuming the preexisting unity and coherence of these objects for the actors involved. Second, for the purposes of my study, the phenomenological approach is useful to grasp how both 'scientific' and 'indigenous' knowledge hold together in certain instances and fall apart in others, depending on how participating actors constitute the corresponding ontologies into acts (or not). In this way, I am able to tend to open-ended ways of knowing in making medicine meaningful and beneficial in a broad sense (beyond the biological). It is this 'space' that I aim to inhabit, one 'in-the-world' at the interstice between foreseen historical possibilities and constituted acts or, in other words, in movement. In this respect, I 'did fieldwork' in both 'scientific' and 'indigenous' ways of making medicine as they aimed to learn from each other at their interstices, or at least converged in a common project. Moving into a phenomenological approach helps find ways to engage conversation between different ways of making medicine 'work' in-the-world.

Taking 'objects' as things continuously emerging through entanglements with perception strays from the causality found in controlled laboratory empiricism, which has the very intention of making an object, as discussed in the previous chapter. In a similar approach, bodies, according to Haraway, 'are not born: they are made' (1993: 372); they are 'material-semiotic generative nodes'. Their bound-aries materialize in social interaction: '"objects" like bodies do not pre-exist as such' (ibid.: 375). Mol (2002) further expresses the 'body multiple', and I follow this proposition implicitly, since it does not add new forms of bodies to a 'core universal biological body', but opens the body-object to the world. As with bodies, I contend that medicine does not exist in and of itself, but is only made to appear as such in relation to multiple situated practices in the inhabited world. These authors, however, generally do not account for their own bodies, which I consider inescapable.

I apply this approach to 'medicine' (plant, bark, pill) because it is not an 'object' to begin with, awaiting to be 'discovered', yet is made into such a thing in particular times and places; further, how medicine is made into a 'thing' greatly differs in terms of its borders, materialities, usefulness and the ways people enter into relations with it (or not). Medicine, in both its 'thing-ness' and relations with humans, is thus continuously being 'done', or 'made' into something that counters a particular physiological process, heals or enhances healing possibilities in one way or another. Ways people enter into relations with things implicate one's possibility to be conscious of this thing, one's ability to see and feel things in certain ways and not in others. For instance, I have been working on the topic of the en-counter between indigenous and biomedical forms of knowledge for almost twenty years, and I am not aware of the same things within the process of making medicine as I was previously. While I have forgotten some things, I have also become aware of new things and possibilities. It was, for example, only in 2007 that I was able to re-ally grasp the centrality of sounds and dance within the indigenous healing practices I studied.

Looking back to shamanic healing practices in which I partici-pated in 1998 and in 2000 amongst the indigenous Madija-Kulina populations of the Brazilian Amazon, sounds, movement and music was also prevalent; however, this had completely escaped me at the time (Laplante 2004). A healing event involved hours of singing by the women before the shamans came out of the forests with their straw masks, and when they did come out, it was because communi-cation had been established with the forest, plants and cosmos. An evening organized on my behalf as I was sick with the flu began with tobacco fumigations in the afternoon and continued with hours of chanting; around ten shamans and myself sat on a fallen tree trunk all facing the forest as we took the hallucinogenic drink *rami* (*Banisteriopsis caapi*) under the meticulous care of one of the shamans, taking our pulse to make sure we consumed the proper amount. My ability to listen to the sounds they were making at the time was certainly limited, since it was only later that I became aware of their central role in the healing performance; certainly, awareness or ability to listen (or not) plays into the healing possibilities as well. My point in telling these events is to state that what I am able to grasp at a particular moment in time both precedes and participates in the 'effect'; in other words, there is no stimuli 'out there' to which homogenous living beings can react as assumed in an experimental setting, rather, there is only stimuli we are able to perceive (or not). This applies to a singled-out plant or molecule and is precisely what the preclinical trial disregards, supposing all biological bodies of a particular species, of the same age and gender, are the same to begin with (only a nuance is made in the different phases of the RCT between infected and noninfected individuals, referring to the disease to be treated by the medicine).

In Merleau-Ponty's idea of continuous entanglements between seeing and moving (1945), prior experiences of seeing or knowing are always fed back through movement, making it so that we are always engaging in the world or medium and with things in new ways. To know *A. afra* in practice thus involves engaging with it; one can engage with it as kin in the everyday or as an external object to manipulate and transform in controlled environments. 'Doing fieldwork' is engaging in the multiplicity of 'reality' in the everyday. Following what is going on in making a medicine into a 'thing' is a particular kind of fieldwork. In this fieldwork I am not an 'observer' nor a 'participant observer', but rather an apprentice or someone sharing a matter of concern with ways of healing from a particular anthropological stance and with particular abilities and not others. My engaged anthropological enquiry is done through attuning my attention and acquiring skills to know medicine. While I developed some listening skills in working with the healers, some anthropological concepts might help explaining what this implies. As such, the three enmeshed themes of embodiment, sentience and mediums opening to 'worlds of becoming' all help bring me closer to 'what was going on' in practice with the actors involved in the preclinical moment, both humans and nonhumans. *A. afra, umhlonyane, wilde-als,* or however the plant being tested is named along the way is also found to know, act, feel, live in mediums and take part in life-making processes with humans in-the-world; thus, its previous experiences within life-making processes will also play a role in its engagements with humans. With this I aim to keep a situated positioning in following what is going on, doing what Kusenbath (2003: 455) calls a 'go-along' ethnography, keeping research open-ended. I aim to learn in engaging with people and plants, even dissipating the ready-made methods of the anthropological discipline such as interviewing, in order to maximize 'the transcendent and reflexive aspects of lived experience in situ' (ibid.) and in vivo.

The phenomenological stance I apply transcends the distance between subject and object, focusing on the engagements in the world. Hence, I am both subject and object, tracing how a wild bush becomes involved in a transnational scientific meshwork[2], recalling ancestral traditions through modern ones in an increasingly large assemblage of people and things. In this approach I also keep in mind Merleau-Ponty's insistence on permeating one's texts. Merleau-Ponty unveils this approach in his way of writing with eye and mind, as a painter who brings 'his body' to his masterpiece. He shows how to respond to his own critique that 'science manipulates things and gives up living in them' (1945) by not doing so himself in writing. To avoid giving up living in things implies keeping some of the opacity of the world. It also invites keeping the world 'alive' and in continuous motion as people and things move with it. 'The animacy of the lifeworld, in short, is not the result of an infusion of spirit into substance, or of agency into materiality, but is rather ontologically prior to their differentiation' (Ingold 2006: 10). Life is then to be understood as 'continuous birth' (Wemindji Cree native hunter, in Scott 1989: 195; see also Ingold 2000: 11) or 'co-naissance' (co-birth, also meaning 'knowing' in French). Along a similar line, Devisch explains how for the Yaka in the southwest of Zaire, who practice a form of *ngoma,* as do the Xhosa healers in the Cape, 'the healing art is a very practical method of intertwining the body with the group and the life-world' (1993: 265). To be able to partake in this 'co-naissance' or knowledge with the actors involved in the preclinical trial, I had to develop new affinities or skills to grasp the multiple ways of engaging with medicine in lifeworlds as well as tie these into ways of writing about these practices within anthropology. Developing these skills involves entangled mind and body, neither preceding the other. It thus involves moving beyond the idea of a closed universal biological body.

Embodiment

In dismantling the reductionist universal biological body upon which biomedicine has come to rely, medical anthropology is 'potentially perhaps the most radical, because this field tackles the Cartesian paradigm at source – in the body itself' (Herzfeld 2010: 432). While subtleties with regard to the compatibility of organs and blood between bodies has become well-known, '[i]t is still commonly assumed in the medical sciences that the human body is readily standardizable by means of systematic assessments, bringing about a further assumption that the material make-up of the body is, for all intents and purposes, universal' (Lock and Nguyen 2010: 20). The body as such, closed upon itself and to the world, passively awaiting the proper stimuli, exists only in the imagination of the experiment and as strongly upheld within the RCT. The clinical trial procedure maintains a universal biological body in its design, one upon which a sole molecule can be tested for its efficacy, notwithstanding its prior and current engagements in the world. The plant's body, from which a molecule or preparation is extrapolated, its place of growth and relations with humans, are also of no concern whatsoever. What are left are temporary objects devoid of any worldly engagements.

Medical anthropologists have largely contributed in attempts to move beyond this 'body proper', essentially showing the multiplicities of bodies in the world. Lock and Farquhar (2007) have regrouped numerous such worthy contributions in their recent anthology, including Mauss's pioneering work on body techniques (1934), as well as an extract from Merleau-Ponty's *Phenomenology of Perception* (1962), somehow excluding Foucault from their account, however. Foucault's groundbreaking contribution to understanding bodies as objects of discipline and normalization (1976) was followed by works on technologies of the self (1988), which move beyond the 'body proper' or towards forms of bodily cultivation. This shift in Foucault's work illustrates ways bodies were put on the scholarly agenda in a two-part motion going from the body to the moving body (Chao

2009). Farnell and Varela (2008) express this more dramatically as two somatic revolutions, suggesting however that the semiotic can be somatic rather than in opposition. The first somatic revolution they explain as the turn towards embodiment such as in the works of La-koff and Johnson (1980), Jackson (1983) and Csordas (1990) while the second somatic revolution corresponds to their proposed notion of 'dynamic embodiment', moving body or to Ingold's (2000) 'dwell-ing perspective' which all embrace further Merleau-Ponty's sugges-tion to understand the body not in terms of an 'I think' yet rather as an 'I can' (1964: 13). Within these two moments in the anthropology of the body seems to be a parallel ground in which numerous medi-cal anthropologists prefer to remain.

Scheper-Hughes and Lock (1990) have, for instance, followed up on the notion of multiple bodies introduced by Douglas (1973; two bodies) and O'Neil (1985; five bodies) by proposing a body politic, a social body and an individual body that are all related through emotions. With the link through emotions, some of the issues with 'adding' layers (here the political and the social) are dealt with, yet the 'individual body' remains unproblematized. Others, such as B. Turner (1992), bring agency back into the picture, in that it is the asymmetry of the objective body and the subjective experiences of the asymmetries of the body that are constant reminders of the co-production of bodies. Lock and Nguyen (2010: 1) address this issue as well in proposing a notion of 'local biologies' to account for biolog-ical and social life as mutually constitutive. In these entanglements, the biological and the social domains, however, remain separate, in the end holding on to an empiricist stance that will attest to this bi-ological difference. The anthropology of biopolitics (Nguyen 2004; Sunder Rajan 2006; Rose 2007) similarly mostly attends to power/ knowledge relations, with perhaps only Fassin (2000) and Leibing (2004) attending to the ways these are lived, at least by others. Gen-erally speaking, these anthropologists have mostly omitted to take their own bodies into consideration.

According to Hsu, while medical anthropology did build on Mer-leau-Ponty's insight of the immediacy between the self and its environ-ment, it did so only to the extent that it takes the body as a starting point for any exploration of the world; 'rather than focusing on the body-as-directly-related-to-the-world, as given in Merleau-Ponty's phe-nomenology, medical anthropologists have foregrounded the self and made the self-as-individual-body, detached from its environment, into a topic of research, much the same as it is in biomedicine' (Hsu 2010: 24). Hsu thus states that medical anthropology has mostly failed to address the interlacing between humans and things (or medicines) or material culture, 'except for the currents that come together in science and technology studies (STS), but those do not generally address questions relevant to the application of plants in medical practice, nor do they take an interest in how spiritual powers become instantiated in the material world' (ibid.: 23).

The anthropology of medicine has for its part attended to the 'social lives' and biographies of medicine (Whyte et al. 2002), with the anthropology of pharmaceuticals (van der Geest and Whyte 1988; van der Geest et al. 1996) adding meaning and symbolic efficacies, yet leaving physiological efficacy aside. These traditions in their own way thus reify 'medicine' rather than investigate the interlacing between humans and 'medicines' in their trails of becoming. The anthropology of clinical trials has further oriented itself in constructivist approaches borrowed from the sociology of knowledge, entertaining a minor interest in materialities and placing a major emphasis on macrostructuralist analysis (Petryna et al. 2006; Petryna 2007a, 2007b) or on research method rationales (Carthright 2007). These works all contribute in some way to understanding the epistemologies underpinning both local and global forms for knowledge that concern my research. I have, however, limited use for this literature, since it only accounts for sociocultural aspects, usually leaving physiological aspects aside or unquestioned, hence attending more superficially to these continuous entanglements.

It is with phenomenological approaches in medical anthropology that I was able to reach an understanding of indigenous healing entangled with materialities. Hsu (2010), Stroeken (2008, 2012) and Geissler and Prince (2010), amongst others, attend to these intricacies. This approach also appears to be undertaken by a good number of anthropologists who pursue long enough the issue of indigenous healing, which eventually leads to a need to really learn through experience. The phenomenological project, as it was first termed by Husserl in the early twentieth century, is to assess the world through the lived body. In his first lecture in 1910–11, Husserl explains that the 'I' in the natural attitude finds itself at all times at a centre of a surrounding (*Umgebung*). In the second lecture, he explains that in the natural attitude, 'every "I" finds itself as having an organic lived body. The body (*Leib*), for its part is not an "I", but rather a spatial-temporal "thing", around which is arranged a surrounding of things that reaches outward without limits' (Husserl [1910–11] 2006: 1–4). In contrast, the phenomenological approach emphasizes the organic lived body *experience,* dissolving the abstraction of the 'I'. Essentially what Husserl is dissolving is the mind-body dichotomy; minds are always being mediated through lived embodiment (ibid.: 157). Further, the world is not surrounding the body; rather, the organic lived body is a very part of the world.

In a later lecture, Husserl distinguishes how

> it is clearly one thing to investigate nature, to describe and investigate things, causal changes in things, temporal orderings of thing-like objectivities, and it is something completely different to leave alone the whole of nature and, in lieu of it, to describe and investigate the experiences of things in their immanence, to describe and investigate what is found in them, how they hang together, how they are motivated, etc., and especially, how they hang together with judgments, feelings, desires, etc. and how they motivate these – and all this under the auspices of a consistent disengagement of any judgment about the existence of nature. (ibid.: 78)

While it is the second program that interests me, I would rather formulate the proposed 'disengagement' as one of continuous deepened engagement as proposed in Merleau-Ponty's phenomenology of perception. This is perhaps why Merleau-Ponty is better attuned with anthropology; indeed, Csordas put him centre stage within the field (Hsu 2012: 56).

Csordas's (1988) elaboration of a paradigm of embodiment has, as a principle characteristic, the collapse of dualities between mind and body, subject and object, a collapse in line with phenomenological approaches. Csordas (1990: 7) refers to Merleau-Ponty (1962), who elaborated embodiment in the problematic of perception, as well as he refers to Bourdieu (1977, 1984), who situated embodiment in an anthropological discourse of practice. For Merleau-Ponty, in the domain of perception, the principal duality is that of subject-object. The body is a 'setting in relation to the world', and consciousness is the body projecting itself into the world. For Bourdieu, in the domain of practice, the principal duality is that of structure-practice. The socially informed body is the 'principle generating and unifying all practices', and consciousness is a form of strategic calculation fused with a system of objective potentialities. Embodiment is the methodological principle invoked by both authors to collapse these dualities.

Merleau-Ponty lays out his position as a critique of Cartesian empiricism. Empiricism is a theory of knowledge that supposes that evidence can be discovered in experiments; evidence-based medicine (EBM) is the fundamental scientific method that tests theories against 'observations of the natural world'. Empiricism asserts that knowledge comes from sensory experience and evidence, however not from a positioning of being in the world, yet from an externalized positioning about what is observable, measurable and representable by the expert eye. Merleau-Ponty does not reproach empiricism for having taken the natural world as the primary theme of analysis. He reproaches empiricism of speaking of a nature as a sum of stimuli and qualities (1945: 48).

> It is absurd to claim that this nature is the primary object of our perception, even if only intentionally: such a nature is clearly posterior to the experience of cultural objects, or rather, it itself is a cultural object. We will thus also have to rediscover the natural world and its mode of existence, which does not merge with the mode of existence of the scientific object (Merleau-Ponty 2012: 26).

Merleau-Ponty argues that perception is by nature indeterminate – it can never exhaust the possibilities of what it perceives. For Merleau-Ponty, the experience of perceiving in all its richness and indeterminacy is the starting point. 'Our perception ends in objects ...' objects are secondary products of reflective thinking. On the level of perception, we have no objects; we are simply in the world. Perception begins in the body; it starts with 'the body in the world'. The body can be self-objectified on different levels as well as in different ways. We can, for instance, maximize the objectification of the body, or follow bodily techniques and strategies to avoid this objectification, or we can transform this objectification, should it have become harmful, as when objects are solidified through protocols and procedures. As with bodies, medicine is not 'out there to discover', nor is it a closed entity; yet it is made into such a 'thing' within an RCT, while through other mediums it is left open-ended.

Merleau-Ponty's project is to 'coincide with the act of perception and break with the critical attitude' (1962: 238–39) that mistakenly begins with objects. Phenomenology is a descriptive science of existential beginnings, not a science of already-constituted cultural products. If our perception 'ends in objects', the goal of a phenomenological anthropology of perception begins and ends in the midst of arbitrariness and indeterminacy; it constitutes and is constituted continuously. For Merleau-Ponty, the body is in the world from the beginning: 'we must return to the social with which we are in contact by the mere fact of existing, and which we carry about inseparably with us before any objectification' (1962: 362). In other words, we must reach a preobjective positioning, one that does not preclude objects but aims to grasp how they are made to appear or how they can be made to appear otherwise. This is precisely the state that healers aim to reach to begin healing, in a way aiming to 'make' the (dis)eased appear otherwise. It is also from this positioning that I aim to grasp how A. *afra* is made into a medicine (or not), supposing it is not necessarily a medicine to begin with, but also supposing it emerges as such in different forms depending on how it is perceived – whether as a medium, as kin, as a bioresource and/or as a pharmaceutical product, for instance.

In parallel to Merleau-Ponty's goal of moving the study of perception from objects to the process of objectification, Bourdieu's goal is to move beyond analysis of the social fact as opus operatum to analysis of the modus operandi of social life. The principle generating and unifying all practices, for Bourdieu, is the socially informed body, 'with its tastes and distastes, its compulsions and repulsions, with, in a word, all its senses, not only the traditional five senses – which never escape the structuring action of social determinisms – but also the sense of necessity and the sense of duty, the sense of direction and the sense of reality, the sense of balance and the sense of beauty, common sense and the sense of the sacred ... the sense of humour and the sense of absurdity, moral sense and the sense of practicality, and so on' (Bourdieu 1977: 124; see also Csordas 1990: 11). While I adhere to the importance of these multifarious senses to grasp practices, I dissociate from Bourdieu's idea of an all-determining 'principle', or of a 'total repertoire of social practices' necessarily compatible with those conditions forming a system, as they recall a form of pre-existing essentialism. This approach has also moved toward imagining embodiment as a container of memories and of information, which has led some scholars to simply move from mind as the locus of a cultural program to the body as such a locus, such as in Samudra's work *Memory in Our Body* (2008). The notion of embodiment I aim to follow is one that is left open-ended and fluid, not contained and accumulating information, yet one that is continuously emerging in engagements with the environment. While there may be 'conditions of possibility' to begin with, these are only those conditions that can be perceived and that are brought into being at a given time and place, or that are successfully imposed and maintained through actors and as enabled through legal instances, formal documents, procedures and standards. These 'preconditions' are however never done in the same way.

Practice operates at the level of preobjective intersubjectivity (empathy and intuition), and it is useful to adhere to this positioning to assess a practice as well as to grasp how it 'works'. At this level, healers do not 'diagnose' but 'discern'. *Isangoma* healing sessions are an incorporation of divine power. These revelations are presented in a variety of sensory modes. Healers are not drawing on a cognitive list of diseases, but are more likely to visually imagine a part of the body, or experience pain in their own body. There is no cognitive act of 'scanning', either of a list of diseases or of body parts, for the one that 'feels right'; rather, they use techniques of the body that are continuously identified, 'discerned' and improvised in the medium. Scientists also develop skills in the medium; however, scientific enquiry has consciously moved away from the senses 'in life' since its earlier forms of experimentation.

The paradigmatic implications of embodiment extend to how we study perception as such. As Csordas (1990: 35) contends,

> Beginning with the experiments of Rivers (1901) in the Torres Straits expedition, anthropologists have (1) considered perception strictly as a function of cognition, and (seldom with respect to self, emotion, or cultural objects such as supernatural beings; (2) isolated the senses, especially focusing on visual perception, but seldom examining the synthesis and interplay of senses in perceptual life; and (3) focused on contextually abstract experimental tasks, instead of linking the study of perception to that of social practice (cf. Cole and Scribner 1974, Bourguignon 1979).

Within a paradigm of embodiment, analysis would shift from perceptual categories and questions of classification and differentiation to perceptual processes and questions of objectification and attention/apperception (ibid.), asking how cultural objectifications and objectifications of the self are arrived at in the first place. If we begin with the lived world of perceptual phenomena, our bodies are not objects to us – they are part of the perceiving subject. The relation between subject and object thus collapses.

Embodiment might further guide how we can understand healing performances. Csordas (1996) provides the contours of a theory of performance specifically to address the questions raised by healing. In particular, he avoids reductive, descriptive accounts that aim to understand healing practices by attributing efficacy to some global, black box–like psychic mechanism such as trance, placebo, suggestion or catharsis. Wreford (2008) alludes to a similar understanding through her experiences of becoming an *isangoma* healer, avoiding reducing these practices to trance, as I will address in-depth in chapter 6. By grounding these practices in embodiment, or in existential immediacy, it is possible to understand how the healing is real. Csordas mentions how it is a fundamental incorporated 'withness' of divine power (ibid.: 108) that might best explain healing insights and efficacies. He thus suggests we need to move beyond the sequence of action and the organization of text to reach the phenomenology of healing and of being healed. The imagined world thus operates empirically as a convincingly efficacious spiritual world: 'Imagination's efficacy to transform orientation and engagement in the world is reinforced by its close association with memory, especially autobiographical memory (Brewer 1986).' (ibid.: 107) The complementarity of imagination and memory in healing,

> ... consists in that imagination is "thickened" with existential care, while memory is "thinned" by the relative ease of imagination. The specific efficacy within this complementarity lies in the juxtaposition of the divine world of the purely possible and the struggling human world of traumatic autobiographical memory, and in the experiential superimposition of the divine imagination upon human memory in imaginal performance. A further element of efficacy originates in the very embodiment of healing imagery, insofar as the body is the existential ground for efficacy in general. (ibid.)

Csordas explains that these primordial experiences of efficacy, the concrete feeling of bodily efficacy, can implicate the experience of divine power: 'The immediacy of the imaginal world and of memory, of divine presence and causal efficacy, have their common ground in embodiment. The moods and motivations evoked upon this ground in ritual performance are indeed uniquely realistic' (ibid.: 108). Symbolic efficacy, as discussed earlier by Lévi-Strauss (1949), points towards a similar understanding of the powers of symbols in transforming physiological processes. This helps in understanding ways medicines are made to 'work' in this corporeal immediacy, sometimes without their necessary materiality. Both Rastafarian *bossiedoktors* and Xhosa *izangoma* involved in my study heal through deepened immersion in the world, respectively translating visions into everyday performances and intensifying visions through sounds and drumming, in these ways tangibly reorienting life in a desired direction. Rastafarian *bossiedoktors* in the Cape live with the presence of Jah (God) within them at all times, to which they refer as a light moving through them, and this is convincing and real when felt. How things are felt are thus what may make the difference between a successful therapy and an unsuccessful one; this surely plays into the ways a body will react (or not) to the same molecule.

I might here refer to another work by Csordas (2007) to clarify this positioning further. Csordas (ibid.) discusses the importance of paying attention to the shadow side of doing fieldwork, which often has 'telling moments'. He tells of two such moments in his own work that had the effect of changing his existential form and interpretive frame, or what he calls a transmutation of sensibilities. One of the stories told follows a Navajo peyote healing meeting in which he felt a spontaneous moment of empathy and intuition, which he finds fair to call a revelation. His example is meaningful to my topic because it involves shamanistic medicine. The intake of the medicine is part of the enhancement of the experience, more particularly when it is consumed when immersed within a habitus (in a Bourdieuian sense). Upon being able to make sense of the dreams evoked by the intake of the medicine within the charismatic habitus, a Navajo healer told Csordas, 'Now you know for yourself how this medicine works. This is your own story that you can tell' (ibid.: 114). Without referring to a necessary preexisting habitus but to an always emerging one, this is precisely how Ingold begins his 2013 book: it is important to 'know for yourself', as he learned to do with Saami people in northeastern Finland. In this way, without clinging on to the assumption of a preexisting habitus and thus adhering to a more dynamic and open-ended notion of embodiment, I aimed to know for myself, moving through the different spaces where medicine was being done.

Sentience

> Rhythm, dance, chant, and melody give the body over to the senses and the life-world.

– Rene Devisch, Weaving the Threads of Life

'Indigenous' healing began to 'make sense' once I started to pay attention to the senses. This was a turning point during my research. It was just such a moment of spontaneous empathy and intuition, a sort of revelation, namely what Csordas (2007) refers to as a transmutation of sensibilities and which I prefer to call a 'line of becoming' following Deleuze and Guattari (1980). This change or transformation in perception can be understood as reaching a plateau, as sometimes occurs in learning a language, when disparate information (words and sounds, for example) all of a sudden begins to work together in ways of communicating with others in the world. I would however argue that such a transformation in perception is less dependent upon immersion into a pre-existing *habitus,* since I was rather unfamiliar with Xhosa healing. Rather, such a transformation is closer to serendipity in the sense of finding common grounds based on each other's past and current experiences and which happen to cohere. At the beginning of this chapter I mentioned that when I asked a Xhosa *isangoma* 'How does *umhlonyane* work?', the healer responded by inviting me to a drumming session. To express the 'obvious' answer to my question through this invitation to a drum session, he made a very significant gesture with his hands falling along his navel area to show how knowledge relied on the ability to feel sounds through ones body. This stuck with me and eventually led me towards the importance of sentience in this research. At that moment it all came together; knowledge was to be felt, making intellect alone inappropriate to follow what was going on. Feeling sounds is in fact one of the prime mediums for acquiring skills to heal through medicine in *isangoma* healing. It took those events to turn me towards phenomenological approaches, which also have to do with a broadening of the senses; it took me years to find a way to reconcile this kind of knowledge with research and with a way to write about my research. This in turn made me rethink the possibility of distance from knowledge expressed and lean towards proximity and immediacy. My research went much more smoothly afterwards. I could finally grasp some of the healing processes that were seemingly escaping me, as well as the scientific filter of knowledge in making a medicine.

Fieldwork is an aesthetic experience (Devisch 1993: 39), a meditative vehicle and journey (Tylor 1986: 140). Similarly, research involves bodily cultivation. It involves developing abilities to feel 'with' people, places and things. It is a process enabling me to attune to some of the sensory legitimacies taking place in engaging with medicine, for instance. Most importantly, perhaps, fieldwork involves developing new abilities in addition to those tailored and privileged within academia, as well as reconciling them afterwards into a written and legitimate account. Anthropologist Lesley Green encouraged me to move towards paying attention to music in healing and even as a way to 'represent'. Sentience is the ability to feel or perceive subjectivity, as distinguished from the ability to think (although it can be argued that thinking also needs to be felt to become real). Sentience is precisely what one learns to unlearn when performing an experiment, or rather, during an experiment one solely attunes to specific ways of seeing, namely, through representational mechanisms external to what is lived, felt and visible through those means.

The anthropology of the senses specifically set out to surpass the visual representational tendency linked to science in order to take into consideration not only other 'worldviews' but also other 'ways of being in the world'. Numerous scholars have noted science has developed a particular kind of vision, often to the detriment of developing other senses. McLuhan (1971, 1972) mentions printing as the pivotal technology of communication enhancing the priority of vision as representation, while Arendt (1958) moves this farther back to the microscope (1590), followed by the telescope (1605), all three technologies tending towards a distant kind of vision and to an Archimedean point in the case of the microscope and the telescope. 'The privilege given to vision to the detriment of other sensorial faculties leads to the autonomy of the spread that Cartesian physics will be able to exploit and which also favours the expansion of the limits of the known universe by the 'discovery' of the cartography of new continents. From this point on, mute, odourless and impalpable, nature emptied itself from all life' (Descola 2005: 97). This particular tailoring of the senses led to unilinear, analytical and distancing tendencies that are still prevalent within the sciences, as portrayed by the RCT. This approach seems to prevail even within what Giddens (1994) names 'reflexive modernization' and in new media today, mining the hegemony of vision and largely sensitizing people to auditory and tactile experiences (Howes 1990: 102).

Multisensorial events abound; however, they do not seem to have dislodged the ways of doing science, which continue to proceed with the idea that humans are passive recipients awaiting external stimulation to function and to heal, for instance. A basic assumption I suspect is largely maintained through the limits perceived in RCT protocols.

> Histories of science may be powerfully told as histories of the technologies. These technologies are ways of life, social orders, practices of visualization. Technologies are skilled practices. How to see? Where to see from? What limits to vision? What to see for? Whom to see with? Who gets to have more than one point of view? Who gets blinded? Who wears blinders? Who interprets the visual field? What other sensory power do we wish to cultivate besides vision? (Haraway 1988: 587)

The senses are moral and political positionings, and where their legitimacies are situated is also crucial to take into consideration. According to Merleau-Ponty, 'the animation of the body is not the assembly of one of its parts against another. ... [T]hings and my body are made of the same matter' (1945: 15–16). The body thus creates open-ended meanings, bridging natures and cultures, hence making 'all technique a "technique of the body"' (ibid.: 24), a reference Merleau-Ponty makes to Mauss's (1934) text 'Les techniques du corps.' While refraining from noticing the sensitive reflexivities in these bodily techniques, Mauss did bring attention to embodied ways of moving. Which techniques are cultivated, legitimized and developed can further be understood as moral and political acts. The senses are, of course, always the mediums through which knowledge is acquired in the world; however, which ones we doubt and to which kinds of practices we attribute truth varies greatly. Restricting the senses to five organs, as is commonly the case in Western thought, clearly emphasizes certain senses while entirely dismissing other more subtle sensory organs as well as their synaesthesia; some remain dormant, while others are cultivated according to shared values and skills.

The RCT, for instance, clearly identifies representational vision as the predominant way of knowing. Part of its explicit procedure is 'blinding', the main sense left to scientists for their assessments of 'true' efficacy. The trained eye is the one who knows. 'Blinding' makes sure that this will not disturb the revelation of 'truth'. It is, however, not the eye per se that is covered, but the knowledge of what is being done, that is, if a person is giving/receiving the 'real' medicine or the placebo. This procedure is not named 'deafening' or 'muting', perhaps because those senses have been disregarded as irrelevant long ago for reasons of 'objectivity'. 'The difficulty of recording smell and taste in some reproducible and reasonably durable medium has marginalized these senses more than any other except touch – which, because it is primarily dyadic and thus relatively private, generally escapes social analysis altogether' (Herzfeld 2010: 432). Further, 'smell, taste and touch' are often considered 'animalistic'. This trend of thought was already present in eighteenth-century aesthetics: 'as long as man is still a savage he enjoys by means of (the) tactile senses (i.e., touch, taste and smell) rather than through the "higher" senses of sight and hearing' (Schiller 1982: 195; see also Herzfeld 2010: 436). The particular mode of sight deemed proper is, however, one of distance and 'objectivity'. This is opposed to seeing receptively, as Ingold (2000) learned to do with Arctic hunters, and to immersion in sound in healing, as I learned to do with Xhosa *izangoma* and Rastafarian *bossiedoktors*.

Stroeken (2008: 466) summarizes the anthropology of the senses as a postcolonial project taking two directions: first, there is the 'antiscopic' approach, challenging the oculocentrism of scientific representation; second, there is the approach that brings subtleties to the quality of vision and challenges its status as exclusively representational (intrusive). Representing the first approach, Howes (1991) and Geurts (2002), following McLuhan (1971, 1972), for instance, emphasize multisensorial experiences in non-Western cultures, basing themselves on the assumption that all cultures are apt to elaborate one of the modes of the senses, that is, vision, tactility, kinesthesia, and so forth, at the expense of the others. Within this antiscopic approach, authors, however, warn against the idea of a simple transition from auditory-oral traditions towards a visual-representational scientific culture (Gouk 2005: 87), or of a simple denunciation of vision to replace it with a new sensibility based on the ear (Erlmann 2005: 5), for instance. Stoller's reflexive exploration of taste (1986) and Classen et al.'s (1994) investigation of smell also reify one of the senses in one way or another, even if it is a 'neglected' sense, or less attended to than vision in the West. In the second approach, as formulated by Ingold (2000), the antiscopic overtones of the first approach are nuanced. Vision can, for instance, be 'representational' and intrusive, as critiqued by the antiscopic approach; however, vision can also be nonrepresentational and receptive, as mentioned above with the example of receptive vision in hunting (Ingold 2000). According to Stroeken (2008: 467), this approach however risks essentializing Western versus non-Western cultures. Stroeken (ibid.) thus proposes to reconcile the two directions taken by the anthropology of the senses in terms of sensorial codes and modes in order to take into account both directions; he also adds that sensorial codes vary similarly according to activities and people involved.

Stroeken's proposal opens the possibility of sensory shifts linked with moments and performances within healing practices; however, the research agenda becomes one of describing what these shifts are and when they emerge, very much as Classen (1999) and Howes (1990, 1991) have done in mapping the senses, albeit with more subtleties. In his recent works, Ingold (2011, 2013) proposes another research agenda, which maintains its emphasis on sentience and moves closer to forms of bodily cultivation. 'Knowing from the inside' is not learning *about* healing practices and the ways senses are evoked and organized; rather, it is learning *from* these ways of sensing in healing. This is a subtle yet definitive difference. In the first case it gives a good description of the ways the senses shift, as in Stroeken's (2008) account of Sukuma healing. In the second case, as I will do here, what one has learned through lived bodily experience from these sensory shifts, or at least those shifts that one was able to feel, is how one thus understands these practices, sometimes also shifting one's understanding of everything else. In this way, paying attention to the senses is not made into an object of study in itself, instead simply becoming a way of doing more attuned research.

I can thus only offer understandings I have acquired through the process of sharing lived experiences with *izangoma* healers, Rastafarian *bossiedoktors* and scientists of other disciplines than my own.

I have, however, not undergone an *isangoma* initiation, as Wreford (2008) has, nor have I become a biomolecular biologist, a Rastafarian *bossiedoktor* or a plant, for that matter. I have, however, attuned my attention to these people and plants to learn how medicine is being done in the different mediums in which they engage. Fine-tuning my own senses within the ways a continuously emerging habitus was being done (or undone) was my way forward. Paying attention to how some lived bodily skills became of prime importance in certain instances and not in others requires developing corresponding skills. These skills are mainly learned through moving through the different mediums.

Mediums

Merleau-Ponty follows Jacob von Uexküll's (1937) notion of *Umwelt*, which he differentiates from *Umgebung* and *Welt*; *Welt* is the universe of science, *Umgebung* is the banal geographical environment and *Umwelt* designates the milieu of behaviour proper to a certain organism, 'that is to say, the ordinary world of perspective and pragmatic experience' (Canguilhem 2008: 112). *Umwelt*, or medium, is thus a qualitative space corresponding to an 'environment of behaviour', a specificity of the living, a world environment, not to be confused with a 'geographical environment', which is in a sense an environment that is delimited from the outside. The notion of 'world' refers to a determined medium in which the organism is invariably engaged. Human life would occur in an infinity of possible mediums; this is the case for animals and plants as well. This is demonstrated most tellingly by von Uexküll's example of the *Umwelt* of the tick, showing how the tick can remain insensitive to all external stimuli from a milieu such as the forest for up to eighteen years, only to release its movement upon smelling the odour of rancid butter emanating from the passage of a mammal, which it then drops onto in order to find blood and perform the tick's reproductive functions.[3] What this example shows is that a reaction to stimuli is always a function of the opening of a sense to stimulation.

Taking this discussion to the experiment, it calls into question the assumption that an animal can adjust to a laboratory setting (seven days are given to the white albino mice in the Ntutela et al. 2009 study discussed in the previous chapter) and thus react accordingly to the sent 'information' or molecule. This question extends to humans, who are also enlivened to certain stimuli rather than others depending on their own engagements in mediums. How an animal or a human can adjust to the controlled medium (or not) comes into play. It can further implicate the plant, which, taken out of its medium and brought into a new controlled one, may not necessarily produce the same chemical compositions as found in known mediums; the plant also adjusts to a new medium. This in turn may mean that an animal or a human may modify their bodily receptions in adjusting to a new medium. Taking the medium into consideration thus opens one to perhaps more grounded and meaningful ways of doing medicine without imposing on the plant, animal or human a situation that in fact takes away its agency within life-making processes.

> The situation of a living being commanded from the outside by the milieu is what Goldstein considers the archetype of a catastrophic situation. And that is the situation of the living in a laboratory. The relations between the living and the milieu as they are studied experimentally, objectively, are, among all possible relations, those that make the least sense biologically; they are pathological relations. (Canguilhem 2008: 113)

For Goldstein (1995: 388), 'meaning' and 'being' are the same. It can thus make a certain amount of sense to study ways to deal with pathology in the laboratory; however, we learn very little about how life radiates in the process. To do so one needs to thus develop sensibilities as a living being rather than disband these as obtrusions to understanding the real.

For my research I travelled through a variety of mediums in following the trails and trials of *A. afra,* and part of my assessment is to take these mediums into consideration, namely, to take into account the important distinction between open-air and enclosed mediums, and to express what it felt like to move through them. 'While the philosophical positivism of the nineteenth century of Comte and Taine insists on the determining side of the medium, making man a simple "product of the medium" (a tradition that behaviourist biology will inherit), Merleau-Ponty brings us back to Lamarck's dynamic intuition' (Alloa 2008: 30). Lamarck's evolutionism is an attempt to hypothesize a dynamic and temporal relation between organism and medium. The sociological perspective that man is, before all things, a being within others is then brought back to its most primordial dimension, which signifies that man is not only an organism/body 'in a medium' but is also an organism/body from and through a medium (ibid.).

Following this intuition, Merleau-Ponty postulates that if there is a quantitative relation between environment and physical objects, there exists, between the medium and their organisms, a qualitative relation. Agamben similarly mentions that 'the forest as an objectively determined medium does not exist: what exists is the forest-for-the hunter, the forest-for the botanist…' (2006: 68). This is the quality that is lacking in approaches that transpose mechanics on the living, missing the dialectic between the living and its medium, and, in turn, lacking a philosophy of life. Mechanism postulates immediacy (immediate action of the perceived on the one perceiving, or the theory of reflex) yet lacks, according to Merleau-Ponty (1945), an authentic thought of perception, since mechanism lacks the fundamental element, 'the living body', which is placed in the centre of the phenomenology of perception. How I feel in a laboratory may be quite different from how my neighbour feels. How I feel enlivened by winds, which another may find annoying, is bound to my living body. How I feel a medicine taken from my backyard, or from a pharmacy, or that is activated by a healer may also vary from how my neighbour feels, because we are living bodies engaged in mediums in different ways and this has 'real' implications in efficacies and healing processes. Engaging in mediums meaningfully further opens onto 'worlds of becoming'.

Worlds of Becoming

The notion of 'worlds of becoming' I want to put forth borrows from Merleau-Ponty's notion of 'world' and anthropologist Tim Ingold's notion of 'trails of becoming'[4]. Both these ideas point towards an understanding of life as a movement of opening as opposed to one of preexisting closure. They invite one to understand the inhabited world as sentient. They are phenomenological in the sense that they are interested in the ways experiences or 'worlds' are made to appear. Finally, they invite one to understand practices through a positioning as a sentient being immersed in the world, which has great implications both in terms of research and in terms of the ways of writing research. It is a positioning that sets itself in direct opposition to positivist science, which makes the researcher separate from what he or she is studying, a positioning initially agreed upon by most of the scientists involved in the preclinical trial I followed, although challenged by its aim to recognize indigenous medicine, which involves more than an active molecule. I argue that indigenous medicine attends to 'worlds of becoming' to know and heal with medicine, while the preclinical trial model attends to a 'one world' or 'one nature' idea, thus creating the tensions playing into the preclinical trial of an indigenous medicine. I will tease out both notions of world as a way to understand what is going on in the preclinical trial of *A. afra*.

The One World–One Medicine–One Health initiative discussed in the introduction has emerged as animal biological studies are joining human biological studies, hence broadening the idea of 'one world' to more living species than solely humans. As the World Health Organization (WHO), the United Nation's Food and Agriculture Organization (FAO) and the World Organisation for Animal Health (OIE) and others unite in the One Health Initiative, hope rests on oneness, as organizations previously tending to humans, food and animals separately join forces. While nonhuman species and food may join humans in this health endeavour, they do so through their externalization from the world or 'nature'. The experts joining the One Health Initiative all implicitly agree upon a particular notion of world that is external to humans (and now to food and animals) and can be referred to as 'nature'. The notion of 'one world' fits into a particular ontology that Descola (2005) calls a naturalist ontology, namely, Western scientific thought, which he contrasts with analogism, totemism and animism. Descola describes the naturalist ontology as the most recent (emerging in the seventeenth century) and unique in its ontological divide between nature and culture. Laboratory science, experimental science and clinical science as we find it today is sturdily anchored in this ontological divide, which supposes a background of 'one nature' that constitutes the 'real' and that is set as hierarchically primary to the felt and the lived (culture), as discussed in the previous chapter. In this ontology, the 'real' or 'world' is thought to be separable from the lived and the felt, something we may progressively 'know' from an externalized standpoint, an observational distance assured by self-effacement as well as rigorous observation in highly controlled environments (laboratories). These conditions are supposed to give access to the way the 'world' works.

It suggests that it is by extracting ourselves from this 'world' that we ought to 'know it'.

Environmental studies may be the most illustrative of this positioning, as they foreground a background of nature from which we as humans are excluded and that we must preserve or conserve. A similar rationale is put into effect when medicine is to be extracted from this environment in order to provide health to humans. These practices rely upon such divisions to proceed with making medicine; the controlled laboratory environment becomes the way to know how medicine might work best for health issues. In the preclinical trial, it is the ontological divide between nature and culture that makes it possible to extract data, parts of the 'real' world, in order to test their effects on humans in an experiment. It is also this divide that has led to the exclusion of the lived and the felt, and their banishment into the category known as 'placebo'. These scientific practices thus suppose the 'world' as external to living beings, which we might 'know' from a distance, as well as test some of its parts, a molecule, for instance, on human bodies. In this positioning, a part of nature is being tested on a human generally perceived as passive and awaiting stimuli, thus relying upon a particular notion of perception. While this is not a problem in itself, it becomes a problem when it claims to recognize medicine known 'in life', such as is the case with most indigenous medicine. The clinical model is unable to take these intricacies into consideration, as is the case within the One Health Initiative as a whole.

Studying the preclinical trial of an indigenous medicine anthropologically, I have come to adhere to another notion of perception than the one enacted in the experiment or preclinical trial, namely, one in which movement between eye and mind is continuous, as proposed by Merleau-Ponty and discussed above and in the previous chapter. This positioning brings another notion of 'world' into being, one that is not a background space upon which cultural diversity can profile itself, but a world of perception, of tastes, of feelings, of life that is continuously emerging. This is what I was invited to delve into with Xhosa *izangoma* involved in the preclinical trial; I was invited to feel the world by assisting drumming sessions. *Isangoma* practices are ongoing in the Cape's townships. They require developing skills to listen and feel, enabling connections with plants, people and ancestors. To learn how *izangoma* heal with plants, I was thus invited to feel with them, deepening connections to places where people live and die. In other words, I was invited to attune my attention, or to know, through deepened engagements in the world. In this way, knowing is movement done in the flow of life-making processes. There is no separate world of nature from which to extract data to analyze or map, nor is there culture to filter out of the experiment. We are always in the world, and knowing from this positioning can or cannot be acknowledged and refined.

Merleau-Ponty's 'world of perception' begins with the notion of *Umwelt* (medium), as discussed above. Tending to the living body takes us from medium to world. Of our body, we have various experiences. We make experiences 'from the inside' as a body that we act and through which we act. We also make experiences 'from the outside' in the experience of being seen. This inherent duplicity of the body was one of Husserl's first intuitions. He distinguished – the German language having two words where Latin languages only have one – between body-object (*Körper*) and body-subject (*Leib*) or living body, *Leib* having the same root as 'life' (*Leben*) (Alloa 2008: 31). Phenomenology of perception is a response to a still too abstract configuration that does not take into consideration that to have a body is not to act on it, but through it. Having a body is then for a living being to join a defined medium. As a potentiality in a medium, the body does not let itself be reduced to a total autonomy of a pure subject, nor to the heteronomy of an environment. Applying this approach to the practices in the preclinical trial requires moving through mediums, which holds the birth of new possibilities of engagements – the emergence of a world.

Merleau-Ponty, and perhaps the phenomenological project as a whole, expresses his disagreement not with the object, but with the methods, of science; he suggests to first awaken to the experience of the world, of which science is solely a secondary expression. Primacy of perception starts from the experience itself, by placing ourselves within the subject. The world and the sense of self are emergent phenomena in an ongoing 'becoming'; I do not see the world in terms of its exterior envelop; 'I live it from the inside; I am immersed in it. After all, the world is around me, not in front of me' (Merleau-Ponty 1962: 42). We are immersed in the fluxes of the medium as Ingold (2011) would say. Through involvement in the world – being in-the-world – the perceiver tacitly experiences as many perspectives as possible regarding an object, coming from all the surrounding things in its environment. This is precisely what I have done, as an anthropologist, with the 'object' medicine. Applying this approach to the plant can also explain its variability as it is found through different mediums; it can help in understanding how it is given the attribute of 'life' in indigenous healing as it is given agency. The plant also acts through a medium, opening to worlds of becoming.

While Heidegger finds closure and encircling in the notion of medium, Merleau-Ponty finds in the medium his conception of opening; both humans and nonhumans are found to be 'situated in' and 'open' to the medium rather than constrained by it. 'In keeping a distinction between animal and human, Merleau-Ponty explains the development of the latter from the former; in utilizing the possibility of the condition of opening, man liberates himself of his objective determinacy' (Alloa 2008: 34). The medium is hence not determining, but holds the birth of new possibilities of engaging through it. These ideas will give birth to the theory of the body as a medium, 'no longer body fusion with an *Umwelt* [medium], yet body as a means or an occasion of the projection of a *Welt* [world]' (Merleau-Ponty 1995: 284).

This leads to the theory of creative expression. Placing ourselves within the subject is then leaving to the body a certain thickness that is lost in the strange, sealed alliance between the naturalism of science and the spiritualism of the object. In the phenomenology of perception, the notion of 'medium' is redefined with relation to the body: the body is no longer 'the transparent envelop of the spirit' (ibid. 1945: 187), rather it is a 'means' (ibid.: 144) to make a medium into a world. Both the animal and the human kingdoms are fundamentally situated in, and open upon, a medium. How we feel as we move through these created worlds varies greatly, as well as how we make others feel as we move through them together. The medium is not constraint, nor predetermining; rather, it is continuous new possibilities of engagement. This opening on a medium and into worlds of becoming can be further understood within the reflexive turn in anthropology.

The reflexive turn in the 1980s in anthropology is linked with what some have called a 'crisis of representation' (Clifford and Marcus 1986), bringing into question one's own tradition of work and one's own tools of research and orientation, as well as one's experiences with regard to possibilities of 'representing'. The reflexive turn questioned whether 'representation' should even be possible, whether it is beneficial, and for whom. Beck et al. (1994) named this moment of the world 'reflexive modernization' – reflexivity as the defining thematic of the second modernity. The particularities of reflexivities can be summed up by nonlinearity, indeterminacies, intentionalities and moving subject-objects. In earlier research, I applied what I qualified as reflexive anthropology – both a method and a research technique (Laplante 2004). Within applied reflexive anthropology, I defined *experiential reflexivity* as prior and actual experience at play in the particular research; *institutional reflexivity* as both orienting topic and a way of writing academically, as also defined by Bourdieu and Wacquant (1992); and *reflexivity of knowledge* as both a way to conduct research through dialogue and negotiation, both shaping and shaped by the people and context, as well as a way of learning about the topic of research (Laplante 2004: 263). Essentially, all forms of reflexivity refer to practices consciously oriented towards grasping the subtleties of what is going on during the research process as well as towards bettering oneself in ways of engaging with people in the world. In this research I add more subtle forms of sensitive reflexivities that one can tailor (or not), akin to bodily cultivation.

In asking the question 'How does the relationship between the mind and the body come to be through cultivation?', Yu (2009: 4) points out how Asian traditions do not assume a sharp distinction between mind and body, while the West has broadly aimed to understand the relationship between one and the other, developing particular relational theories, 'including parallelism (mental and somatic events accidentally coinciding without causal relation), reductionism (mental events are actually bodily events or vice versa), and interactionism (bodily states both influence and are influenced by mental states, or one state affects the other but not vice versa)' (ibid.). This way of framing issues through an assumed dichotomy of mind and body is important to acknowledge in my discussion, since, as Yu similarly notes in Asian traditions, the key precept in Xhosa and Rastafarian healing practices (albeit different) is the need to achieve (rather than discover) unity in mind and body. For the Rastafarian, (dis)ease is due to incoherence between words and acts, while the Xhosa healers orient their healing sessions through one and the other indistinctively. This might point to some discrepancies in adopting a research method that prioritizes other styles of bodily cultivation, hence the need to open myself to these approaches in this research.

To remain open to sensitive reflexivities becomes central to research as a never-ending open process always in emergence. Merleau-Ponty (1964) brings attention to the body as already operative and current, always already there as an articulation of seeing and moving that is never submitted to scientific determinism. 'The eye inhabits the body as a man inhabits the world' (ibid.: 20), and man sees only what he looks at. Extracting things from their living environment is a decision to disregard the living environment as significant. Dismissing 'life' in the researcher or dismissing the presence of the researcher is also a decision to disregard the researcher as significant. These are ontological positionings that cannot be assumed to be universal nor useful for understanding how a people make medicine work through modes of bodily cultivation.

Just as Haraway (1988) states that we are never fully present to ourselves, Merleau-Ponty (1945) reminds us that we never know ourselves completely; my body is a number of things, within things, without simply being a thing. As obvious as this may sound, it is often forgotten within a meshwork of abstractions, and in following procedures and bureaucratic requirements. Reaching into the preontological can maximize the potential for reflexivities. This is my intention in studying the preclinical trial of an indigenous medicine, namely to work more closely and consciously with nonstandardized forms of healing and appealing to varying synaesthesia of sensing. Reflexivities have numerous ramifications in the ways I write as well. Most importantly, I've aimed to mesh academic expectations I imagine in anthropology with life involved in the trial. In particular, I mean to let indigenous knowledge breathe into the text as it has become my experience, even if this is more difficult to achieve, than, for instance, documenting these or any other practices from the outside. Further, I hope to remain open to the medium and world as well as engaged in what I write, as it is a part of my life. Writing is an example of the government of oneself, offering insights on the philosophical cultivation of the self (Das 2003: 96). While I cannot claim as Lévi-Strauss (1971) does to resort to classical music to make the sensitive intelligible and organize the structure of my book (see also Clément 2009), I can claim to resort to other movements to take me into my text. Most of my thoughts are constituted in situ and during trails I follow during walks, cycling and skiing. These thoughts are even more so woven into tending to the lives of my children, my own life and the lives of those around me, including plants and animals. To me, these are all parts of knowing healing and medicine. These experiences are what keep my thoughts alive and flowing.

More specifically, the length of time I have dedicated to attuning my attention to medicine anthropologically has an important role in this approach; engaging in the world is always a new experience, yet it builds on previous experiences and on other simultaneous ongoing experiences, which are all part of my acquired abilities of knowing and not knowing practices. There are, however, 'plateaus', 'surprises' or pivotal moments of experience that can shift all perception. It is upon the latter that I focus, my own pivotal moments of experience and those of others. My methodology is hence about co-birth or knowing of body and world, about the ontological consistence of the sensitive, the ontological power of living and its participation as a dimension of the world. I will not provide a 'representation' or a view from the exterior, but rather a sense of what is going on in life and medicine in the worlds of becoming in which I partook, a reality that cannot be petrified nor objectified.

Latour (2005) offers a route to pay attention to this emerging world in the medium, inviting one to follow actors (both human and nonhuman) as they make the social through their associations with one another. This is always showing the emergence of a world; hence, knowing things becomes about tracing the social in its never-ending emergence, and not being the only one 'conscious' of these traces; rather, it is about letting the actors, including oneself, inform continuously how they create the social as they bring it into being. Beyond Latour, I have borrowed tools from phenomenology, which I have portrayed using the themes of embodiment, sentience and mediums, and I have described how they open onto worlds of becoming. My approach tends to the immeasurable as much as to the measurable; however, in both cases I am not looking to classify, objectify, typify or even 'represent', but to evoke ways of engaging in medicine, how life is attended to, and through which hopes, desires, mediums and worlds of becoming it emerges.

Notes

1. Phase I studies are usually conducted after the safety of the new intervention has been documented in animal research (pre-clinical study); their purpose thus being to document the safety of the intervention in humans. They aim to determine if there are side effects and how the drug is metabolized and excreted. They can take up to one year and imply the participation of between 20 and 80 healthy volunteers. If Phase I shows a success rate of 70 per cent, Phase II can begin. Phase II studies are still about safety but the emphasis is on efficacy. The aim is to obtain preliminary data on whether the drug works in people with a certain disease or condition. Between one hundred and three hundred diseased subjects are to participate in this phase which can take up to two years. For controlled trials, patients receiving the drug are compared with similar patients receiving a different treatment – usually an inactive substance (placebo), or a different drug. A double blind controlled trial is when both investigators and participants do not know who will receive the drug and who will receive the placebo. Phase II moves to Phase III if the success rate of the new drug is 50 per cent or more.

   Phase III are studies about efficacy, some say they are typically effectiveness trials in 'real-world settings'. Most Phase 3 trials are randomized controlled trials (hence controlled randomness which may or may not correspond to 'real-world settings'). They are large-scale studies implicating between 1,000 and 3,000 diseased subjects, and decisions on how to proceed are made through discussion between the Food and Drug Administration (FDA) and the sponsor. This phase can take between two and four years and needs a success rate of 80 per cent with further approval from the FDA afterwards (the FDA usually approves between 25 and 30 per cent of all products). Phase IV is required to complete the RCT and can take between two and ten years and implicates several thousands of diseased subjects. The main purpose of this phase is to review cost benefits and to proceed to post market surveillance requirements to gather more data on safety and efficacy, to identify and monitor possible adverse events not yet documented (Stolberg et al. 2004).

2. 'Meshworks' can be understood as a 'flow of material substance in a space that is topologically fluid' (Ingold 2011: 64), to be differentiated from the notion of 'network' defined by Latour (1987) as knots and nodes connected to one another with links and meshes, transforming the scattered resources of technoscience into a net that might seem to extend everywhere. This might express the difference between a science and technology approach, performed in an externalized fashion, and what I do here in an engaged fashion, namely, through improvising my way through the different mediums.

3. The example is taken up by Canguilhem (2008: 112–13), following von Uexküll (1937: 119).

4. A notion itself borrowed from Deleuze and Guattari's notion of 'lines of becoming' (1980).

Chapter 3
Tracing Medicine
Wayfaring

> Bathed in light, submerged in sound and rapt in feeling, the
> sentient body, at once both perceiver and producer, traces the
> paths of the world's becoming in the very course of contribut-
> ing to its ongoing renewal.
>
> — Tim Ingold, *Being Alive*

Tracing medicine, or following what is 'going on' in making medicine
'work' in the world, is to move through various mediums in which
people are engaged. A specific preclinical trial indicates the trails to
follow, experiencing where they lead. I have thus far explained the
tools and stance I have largely borrowed from phenomenology and
tailored to an anthropological approach. In this chapter I introduce
the way I entered the field, first locating the mediums navigated while
tracing medicine, what it felt like to move through them and how I
was able to grasp their realities. Second I describe how I entered the
preclinical trial of *Artemisia afra* through The International Center
for Indigenous Phytotherapy Studies' (TICIPS) leading experts in
biochemistry. Third I introduce the human and nonhuman actors
I was led to meet along the trails and trials of making *A. afra* into a
biopharmaceutical. I have mentioned in the previous chapter how
human and nonhuman life occurs in an infinity of possible mediums. This was particularly apparent in the context of the preclinical
trial, taking place in mediums ranging from high-tech laboratories in
the United States and in South Africa to the low-tech settings of tin
houses in the townships; the plant was carried along these paths, yet
the plant also guided these paths in varying directions.

To begin, I became acquainted with TICIPS' researchers through email conversations during the fall of 2006. I was based in Germany at the time, just beginning a five-year position as a senior research fellow in the Biomedicine in Africa research group at the Max Planck Institute für etnologische forshung in Halle (Saale), Germany (2006–2010). Through my affiliation with this Institute that, in turn, cooperates closely with The Martin Luther University of Halle-Wittenberg which is nearby, we negotiated a Memorandum of Understanding between the Departments of Anthropology of this University and the University of Western Cape (UWC) in Bellville, South Africa. The UWC became my host institution, namely where most South African TICIPS' researchers were based. Following a tradition established by a partnership between Missouri University (MU) and UWC to partner an American scientist with a South African scientist[1], I was partnered with the biochemist leading the TICIPS's Traditional Healers core.[2]

I pursued preparatory fieldwork in South Africa in January 2007, a longer eight-month stay in Cape Town the following year (September 2007–March 2008), as well as shorter trips to both South Africa and the United States until early 2010. Following TICIPS' trial was not a straightforward path. The times, actors and locations where the process occurs are multiple and spread over two continents, as well as overlap other trails and trials as it aims to attend simultaneously to a world pandemic and to indigenous roots and dignity. The Cape lends itself well to such a process since it is filled with extreme contrasts in livelihoods existing side by side, making it possible to move from one world to another in just a few minutes. Knowing these mediums and their encounters will help us grasp what is going on in and at the peripheries of the preclinical trial of an indigenous medicine that seems to tie them together in a fragile way.

Mediums, People and Plants: Rough Grounds

In its simplest explanation, the shared medium for human and non-human beings is air. 'Of course, we need air to breathe. But also, offering little resistance, it allows us to move about – to do things, make things, and touch things' (Ingold 2008: 23). Mostly unpredictable, air is fluid and in constant movement, making our actions indeterminate or always renewing themselves. Second, humans, at least, walk the world with their feet on the ground. This, humans have in common, although the kinds of shoes we wear, our rhythm of walking and other modes of locomotion we use may greatly modify how we feel the world through our feet (or not).

In practice, my fieldwork was largely conducted by walking to meet with other people and plants. It also happened in car rides from home to various localities in and around Cape Town. In addition, it occurred in the time spent walking to meet with people involved in the preclinical trial. Since I began to follow what was going on in the trial with scientific experts, my first encounters were in university offices, sitting in a chair and discussing, very much like what goes on in my own university office when students and colleagues come to visit. Once in a while we moved around to visit the administration or other colleagues, but most of the exchanges occurred while sitting and talking, sometimes exchanging relevant written works, emails and drawings on a piece of paper to express thoughts. I also attended numerous meetings with healers and scientists, conservationists and city managers, discussions which took place around a table, as in a classroom setting, or even in auditoriums (figure 3.1).

Figure 3.1. Indigenous Knowledge Systems (IKS) branch meeting, November 2007

In fact, A. *afra* remained a word and a 'thing out there' for me for quite a while, without its materiality. While A. *afra* is evoked by scientists in university quarters, it is only 'represented', such as promising molecular configurations of A. *afra* that can be shown on computer screens. There was very little for me to grasp other than the hopes and expectations of the trial. These seemed to make the whole endeavour worthwhile. The grounds were polished and uniform; however, they were slippery and with little meaning. Information was provided with regard to the steps undertaken in research, for instance, issues with one or another procedure or colleague, and most felt restraint working within the confines of the predesigned model. I'll return to the details of these impressions of distant control when introducing the actors in the following section, but I would here like to contrast this with the milieu of the *izangoma,* where the material presence of the plant is equally absent.

I was invited to a variety of settings by the healers in and around Cape Town. As mentioned in the previous chapter, it was being invited to a Xhosa healing drumming session that was most telling of the enhanced connectivity between people and plants in these mediums. This was unique; its corresponding intensity was only found in Rastafarian festivals. Yet in that tin house enclosure, the intimacy of forty people and the proximity to the event, the drumming, the dancing, the dirt floors, the cracks in the walls and perhaps most of all the particular stormy day with rain, wind and hail hitting the tin roof, made for a deeper experience. Wreford (2008: 165) mentions a similar violent storm, a thunderstorm in this case, during her graduation as an *isangoma*; the intensities of the skies confirm or invalidate the success of healing events and are fully embraced as part of the indeterminacies. Cape Town, situated at the tip of the African continent, is known for its wild winds and shipwrecks on the coast. The 'Cape Doctor southeasterly wind' is rumoured to clear the city of pollution, and is perhaps given that name to compensate for how locals mostly find that wind unpleasant and irritating.

I always find strong winds enlivening and capable of transforming the course of events, for instance, during this healing session in the 'open air'. Winds might also affect A. *afra*'s growth. A. *afra* has a wide physical distribution, from South Africa to areas reaching as far north as Ethiopia, growing in varying altitudes, mountains, valleys and fields. A. *afra* was also grown in the sandy backyards of most of the healer's houses in the townships. Since *A. afra* is a common sub-Saharan bush, I also saw it in the 'wild' during my travels, by the side of the road, in grasslands and open woodlands. However, it was nowhere to be found during the healing session. I was later told it had been placed under the bedsheets of the woman to be initiated the night before the session for means of purification. Plants are rooted and can move with the wind, as well as grow and spread through their seeds. Plants' indigeneity can however be uprooted through space and time as well, as Ives (2014) shows for South African rooibos tea's indigenous ecosystem moving southward. Plants also travel through human manipulations of all sorts, which is precisely what is done in a preclinical trial. In my case study, attention is turned towards *A. afra* precisely because humans use it in various ways for healing purposes. I am interested in both how this occurs in the preclinical trial as well as what is done outside the trial, in particular with regard to practices made to appear as indigenous.

In both mediums, the researchers' entry into the micromolecular aspects of *A. afra* and the *isangoma*'s purification manoeuvres connecting with the initiate through *umhlonyane*, lay particular engagements with the plant. Whether it be through visual technologies or mental calculations, *A. afra* is on the researchers' daily agenda as an abstract object broken down into some of its internal workings, yet nevertheless some kind of connection is established between the researcher and the botanical plant entity, which has completely lost its material form as found in the wild, or at least anything recognizable to the untrained eye. Can it be possible this is also the case with the *isangoma*? Without training in either practice I may as well take this 'fair' positioning. I will return to introduce these actors' engagements and disengagements with the plant in more detail below as well as in the following chapters, but first I will tell of the other mediums through which I encountered *A. afra*. Interestingly, *A. afra* is made to appear in its materiality in laboratories, the backyard of people's homes, botanical gardens, markets, shops, valleys, farms and festivals.

Laboratories, Botanical Gardens and Backyards

Laboratories were where I first entered *A. afra*'s physical presence, and implied more walking and looking, but no touching, smelling or tasting. I visited six laboratories throughout the study, some a few times, sometimes by myself, other times accompanied by other visitors; laboratories were sometimes temporarily inhabited by experts and technicians, and they were all the time furnished with machinery, ovens, flasks and various plant preparations. Laboratories are similar to each other as they all control much of the medium, including the air, which is sometimes vacuum-sealed. They, however, differ in the rooms and spaces they occupy as well as the role they play within varying experiments. The South African Herbal Science and Medicine Institute (SAHSMI) laboratory at UWC was filled with students, and the only laboratory I visited that was in an open-ended room that I could walk through and access Internet on one of the computers. All the other laboratories were in closed rooms, usually with controlled access. Since I paid more attention to the transformations of *A. afra* into an experimental form rather than to those of tuberculosis, I'll here describe what it felt like to move through the laboratory in which *A. afra* was being prepared for the TICIPS preclinical trial in UWC locales, offering some comparisons with a high-tech laboratory built and utilized by the Indigenous Knowledge Systems (IKS) branch of the South African Medical Research Council (MRC), the IKS Lead (Health) Programme's laboratories in Delft. No special clothing was required to visit these two laboratories; however, it was understood that nothing was to be touched, while all was accessible to the eye. The liveliest laboratory of the two mentioned was inhabited by a pharmacologist who seemed to have spent a lot of time there.

The pharmacologist James Mukinda from UWC leading the research to establish the toxicity of *A. afra* in mice (and whose publications I discuss in chapter 1) enthusiastically took a student and myself on a long tour of his laboratories, which he obviously deeply cherished as a location to learn and experiment. We visited three different rooms with his lively guidance. A first room was filled with flasks of different solutions, some from his own experiments of *A. afra* mixed with liquids in a flask. The second room was also dedicated to experiments with *A. afra*. A large cotton bag filled with dried *A. afra* was at the back of the room, ready for use (figure 3.2). Our guide provided a gloved demonstration of how he was currently testing *A. afra* ground in animal feed: a paste was prepared with a determined dosage of powdered (figures 3.3 and 3.4) and liquefied (figure 3.5) *A. afra*, which was then mixed with animal feed and

Figure 3.2. TICIPS project A. *afra* powder and powdered animal feed, University of Western Cape (UWC), October 2007

Figure 3.3. Dried leaves of A. *afra*

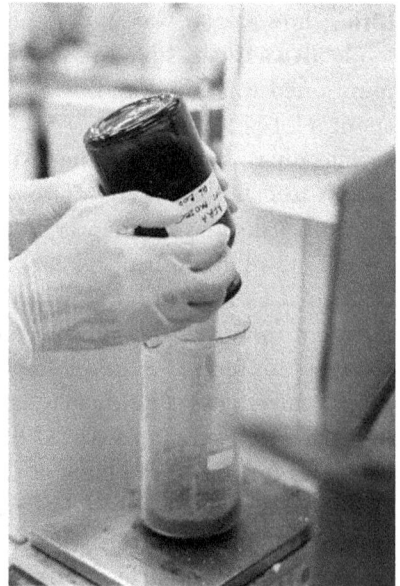

Figure 3.4. A. *afra* powder

filtered through a grinder similar to a food processor, as if making pasta (figure 3.6). His white coat and white gloves were slightly worn down, showing someone working with materials. This second, rather large laboratory room had a small adjacent room. There stayed some twenty white albino mice, part of the experiment to test the level of toxicity of the animal feed with *A. afra*.

Our pharmacologist had obviously worked years in these rooms, which gave the impression that they constituted his 'modest kingdom of knowledge', even partly his home, since he had come from the Congo to study and spent most of his time at the university. The laboratories were good-sized rooms with modest technologies and seemed inhabited and utilized. They played up to the stereotypical image of UWC as the 'coloured' university 'for the people' situated in Bellville, in the periphery of Cape Town. It had noticeably fewer resources than the University of Cape Town (UCT), the international English university situated in a magnificent location at the foot of

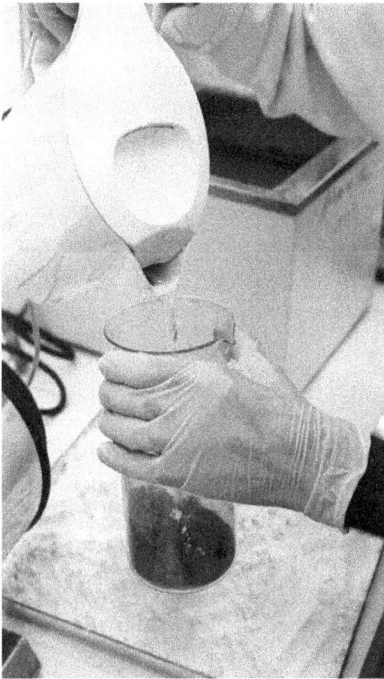

Figure 3.5. Liquifying *A. afra* with boiling water to be mixed with the animal feed

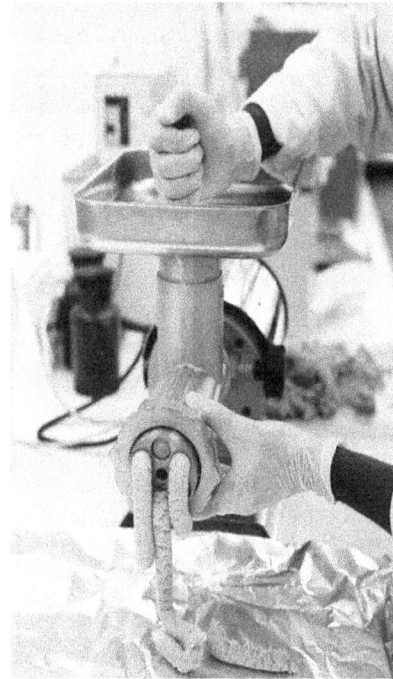

Figure 3.6. Grinding *A. afra* animal feed

Table Mountain, and Stellenbosch University, a more 'Afrikaans' University located about fifty kilometres from Cape Town, where apartheid originated as well as ended. South African political histories are well laid out in the contrasts between these three universities. The collaboration between UWC and MU was highly prized by the researchers of TICIPS. This was also felt in relation to myself, since I did not arrive without affiliation from a renowned institute: the Max Planck Institute had a high notoriety, making my affiliation with TICIPS more welcomed, and my role of anthropologist seemed to be desired.

In contrast to the UWC laboratories, the laboratories in the Delft facility were of an industrial size in a somewhat deserted area on the other side of the highway from Delft township on the outskirts of Cape Town. Newly built with the latest technologies and high-security monitoring, it was a location for experimentation yet also for 'show', with numerous visitors, businessmen, groups of healers, scholars and students passing through on a regular basis. Delft is the MRC's laboratory facility that holds a number of laboratories such as the Primate Unit and Delft Animal Centre, including the IKS lead (Health) Programme's laboratories which most interested me during this research. 'Amid a dozen or so other state-funded laboratories, lies the garden of the new Indigenous Knowledge Systems Lead Programme' (Green 2009: 30). The garden is situated at the entrance of the Delft facility, directly beside the parking space for visitors, thus conveniently situated for visitors to notice upon arrival. I visited the IKS lead (Health) Programme's laboratories as well as the nearby garden for the first time with a group of forty healers in December 2007 (figures 3.7 and 3.8).

In this garden, a number of indigenous plants can be found, amongst them A. afra. A tall wooden fence, surrounded the garden, with some plants identified by their Xhosa, Sotho and Zulu names, others in Afrikaans, but most in English and in Latin. A Xhosa healer commented he had 'helped set this up, but with nothing in return.' While this accomplished isangoma had not received recognition for his work and his abilities in the conception of a 'standard' garden, he mentioned further disappointment with how the laboratory was hiring some initiate izangoma who had not completed their healing training. I found it interesting that some of these young initiate healers had previously presented themselves to me as accomplished

Figure 3.7. Visit to Delft laboratories' medicinal plant garden with a group of *izangoma* and *izinyanga*, November 2007

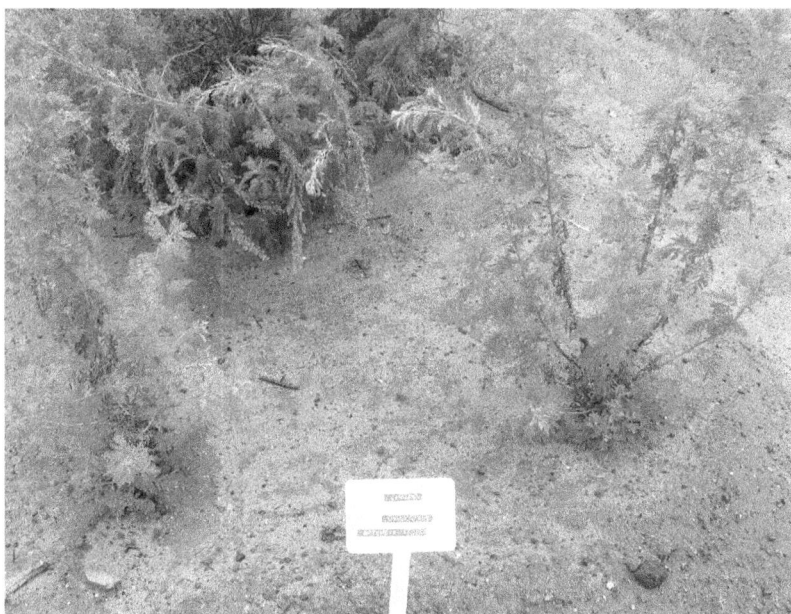

Figure 3.8. *A. afra* on display in Delft laboratories' medicinal plant garden (inscription on panel reads 'Lenga, Umhlonyane, Artemisia Afrika'), November 2007

healers. One of the initiates was a young girl doing her bachelor's degree in medical anthropology, and one of her colleagues had a bachelor's degree in sociology. For these people working in a laboratory, it is likely that academic knowledge in any field seemed better recognized than local experience and abilities acquired in indigenous medicinal knowledge. Further, the apprentice healers were tasked with the organization of traditional healing workshops across the area, making me think of the irony of apprentice healers teaching elderly, accomplished healers about their own trade.

After our visit to the garden in the Delft facility, we entered the building, a very large compound with a series of adjacent rooms similar to a huge warehouse. It had some offices and meeting rooms; most of them were vacant during this visit as well as in subsequent visits. When people were in the rooms, they seemed lost in a very big space, often near very large state of the art technology, which also appeared a bit lost in much more space than was needed. These rooms were vacuum-sealed and impeccably clean. During our visit, laboratory technicians were available to guide us through the laboratory rooms to explain what was done with which machines and which steps transformed a plant into a pill, capsule or powder (figures 3.9 and 3.10).

During a visit in the same laboratory a few months later with my colleagues from the Max Planck Institute's Biomedicine in Africa research group visiting the Cape for our conference, one of my colleagues, an anthropologist, formerly a laboratory technician in the United States, expressed 'doubt' with regard to the 'authenticity' of the laboratory. My own previous experiences in laboratories in Canada were mostly in university settings, like the one at UWC mentioned earlier. In South Africa, the laboratories that most impressed me were those of the Institute of Infectious Disease and Molecular Medicine (IIDMM), part of UCT. This is where tuberculosis samples from South Africans infected with the disease were temporarily kept and prepared for travel to U.S. laboratories. Another laboratory room with infected tuberculosis cells was vacuum-sealed; people entered with heavy garments from behind a window, where we could watch them from the outside. In these laboratories, I met students at work, proceeding to different steps of research, mostly repeating certain actions in verification of procedures for their professors, similarly to what was going on in the pharmacology lab and SAHSMI of UWC

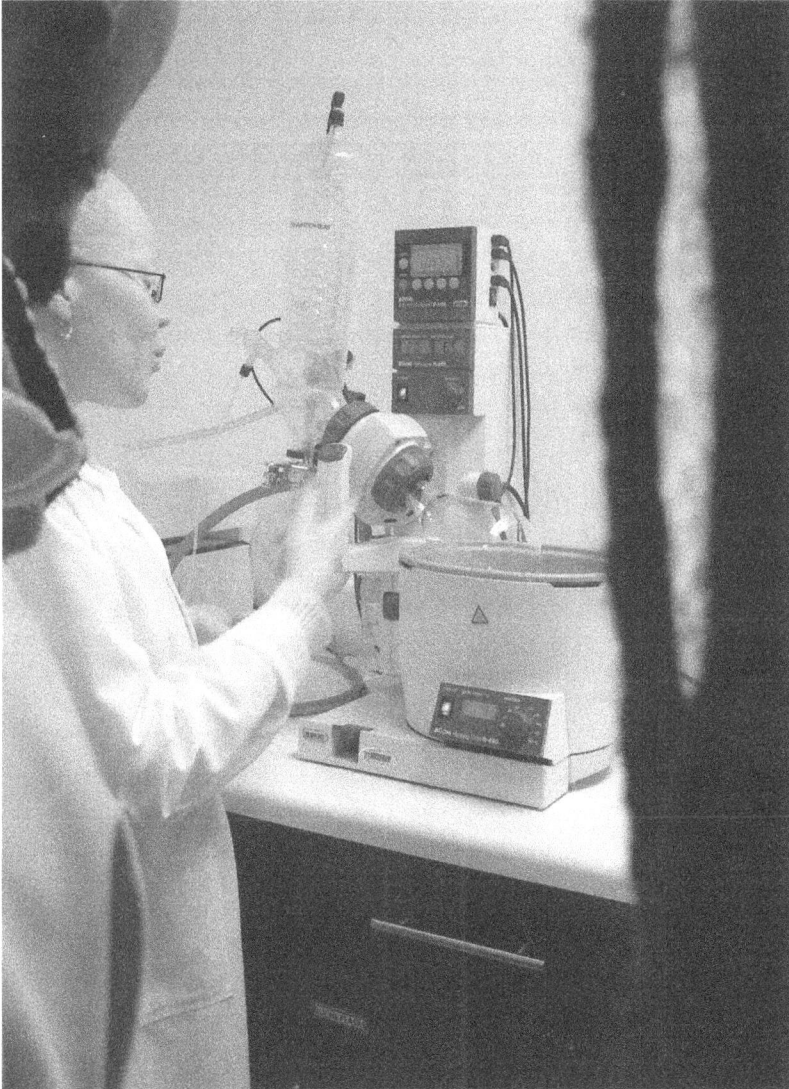

Figure 3.9. Guided tour of Delft laboratories with a group of *izangoma* and *izinyanga*, December 2007

mentioned above. These laboratories were all lively and filled with people at work. The difference between these laboratories and the Delft facility laboratories was the inactivity, perhaps in comparison to its spaciousness; the rooms were huge, with very few impeccably

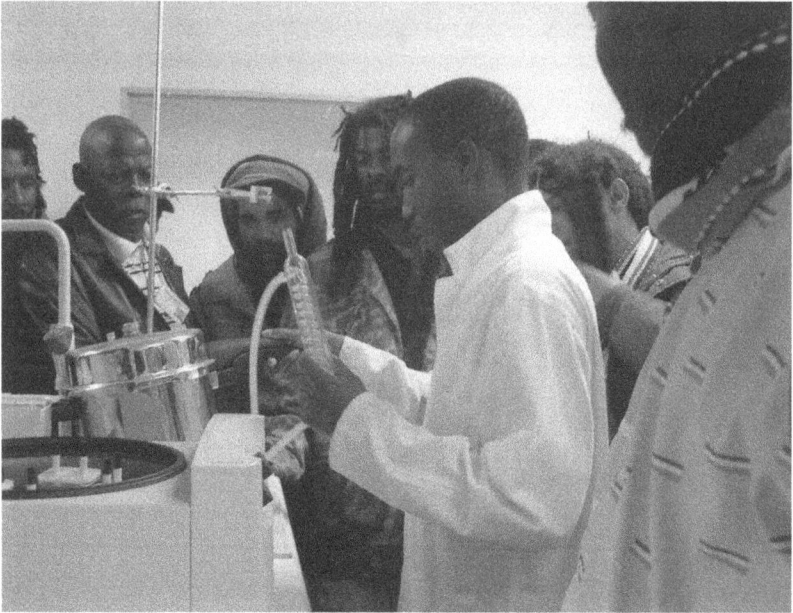

Figure 3.10. Guided tour of Delft laboratories with a group of *izangoma* and *izinyanga*, December 2007

clean pieces of machinery in a corner. Perhaps because we were passing through as 'tourists', it gave the impression of an exhibit, and when a lab technician appeared, it gave the impression of a staged performance, one that was different from laboratories elsewhere, in which we were really 'disturbing' people at work. It is not my purpose here to judge the authenticity of one laboratory or another, but to note how laboratories are fabricated mediums that can be more or less convincing as places where work gets done.

For the healers, the IKS lead (Health) Programme's laboratories in Delft was deceiving, not because other laboratories were more convincing, but simply because the whole setup did not seem to provide the means to understand how indigenous medicine worked. The garden did not seem 'alive' or useful, nor did processing the plant through machinery or testing on animals in cages seem relevant to understand the efficacy of a medicine in and with people. One of the Rastafarian *bossiedoktor* was rather in awe of the sophisticated machinery transforming a plant into powders, capsules and pills; he exclaimed incredulously, 'They have been hiding all of this from us!' Some were impressed with these procedures, while others were disappointed or sceptical either in general or with regard to this particular laboratory. While I have no experience working in laboratories and never got anything 'done' in laboratories, walking through this particular laboratory, I shared the feelings of both the healers and my colleague (albeit for different reasons) that the laboratory did not seem completely 'real'. It was real in the sense of it being there, with all its magnitude, yet it did not seem 'alive' in the sense of a real place to make medicine available to the people who would need them. The Delft facility seemed to serve the purpose of showing businessmen the good investment that could be made there. It felt 'orchestrated' or for 'show', even with livestock such as vervet monkeys in majestic outdoor cages and horses in stalls for experimentation, and complete with lab technicians, secretaries and novice healers. Perhaps it was the feeling of walking through a five-star hotel with very few guests or staff that attested to its loneliness and distance from people and the inhabited world.

On the other side of the highway from the Delft facility was Delft township, where a family of Rastafarian *bossiedoktors* and musicians lived on a tiny parcel of land, eight people sharing a twenty-square-foot cement house. They had a huge bush of *A. afra* growing near the gate of their parcel of land that seemed to be part of the family; part of its stems were dried and kept in an adjacent tin house or in the trunk of the car, ready to sell in the market. *A. afra* was in proximity for use with feverish babies, for tea, for taste, for smell and for sale, available for immediate usage of all different kinds and as part of the family's livelihood. The contrast between the two worlds on each side of the highway was striking, as were the relations with *A. afra*. The laboratory has left the 'living' altogether; even if explicitly dealing with the 'living', it does so in a narrow, disembodied manner. Delft's botanical garden was in the open air, yet fenced and not for human everyday use, making it rather like an exhibit featuring *A. afra*. The *A. afra* in the Rastafarian family's garden was livelihood itself; it was kin, a large bush growing under the clothesline in the backyard, available for everyday use (figure 3.11). Community garden projects were also found in other Rastafarian communities, in keeping with both proximity and with opening to others in the neighbourhood (figure 3.12).

Figure 3.11. *A. afra* bush under the clothesline in a Rastafarian family's backyard, Delft township, Cape Town, February 2008

Figure 3.12. Rastafarian community garden project showing *A. afra* growing and drying in the sun, Muizenberg township, Cape Town, February 2008

While in the two cases above the plant is both grown and used for various healing purposes, Delft's medicinal plant garden is rather for display. This is also the case in Cape Town's internationally re-nowned and stunning Kirstenbosch National Botanical Garden at the foot of Table Mountain, where *A. afra* was on display in six different locations. All six panels first indicate the Latin name of *Artemisia afra,* part of the Asteraceae, or daisy, family. Below the panels provide the plant's English name ('wild wormwood'), its Xhosa name ('umHlonyane') and lastly its Afrikaans name ('Wilde-als'). One of the panels, the only one that adds the qualitative 'African' to the English name ('wild/African wormwood'), uses the Afrikaans name as a reference in its short description: 'Wilde-als brandy is an old remedy for colds, coughs & chest ailments, as well as for heartburn.' This obviously refers to a traditional use of the plant by the colonizers, the ones who brought brandy along. A second and third panel use the English name without the 'wild' in the description: 'wormwood leaves & stems are used to treat constipation & colic, & to destroy intestinal worms'; 'wormwood is added to bath water to wash and soothe wounds, sores, rashes, bites and stings'. This perhaps refers to domesticated forms of *A. afra.* A fourth and fifth panel use 'wild wormwood' in their descriptions: 'wild wormwood is a cleansing, disinfecting herb used as a wash for sores and rashes'; 'wild wormwood is a warming, herb used to treat menstrual chill after childbirth'. Only one panel uses the Xhosa name in its description: 'umHlonyane is used to treat toothache, gum infections and mouth ulcers'. Without overanalyzing the conscious decision making behind the organization of these panels and descriptions, of interest is the difficulty in linking everyday practices with medicine to a particular tradition or 'culture', and perhaps most of all the lack of attention paid to these belongings.

If knowledge is understood as something that lies in experience, the panels clearly exhibit disembodied as well as decontextualized information. Very little is told of *A. afra*'s life with humans in these dry and clinical descriptions of physiological uses; only the 'brandy' and the 'bath' perhaps give hints indicating which people the panels might be referring to (the colonizers). The use of the term 'wild' in some cases and its nonuse in others is also left unexplained. In all cases, it is the same *A. afra* bush on display, or is it? Healers will make distinctions between the plants, as will molecular biologists, albeit in different ways. The healers will make differences with regard to where the plant is to be found and how it should be tended to, while biochemists are aware of each plant's unique chemical constituents depending on their location of growth and the cultivation techniques utilized. The exhibition of *A. afra* in the Kirstenbosch National Botanical Garden seems to make other distinctions, keeping some of its multiple uses and varying its vernacular names; it is somewhere between *A. afra*'s variability in life and its molecular complexities. Somewhat acknowledging this disembodiment of knowing and exhibiting *A. afra* in this way, the Kirstenbosch National Botanical Garden recently added a 'Useful Plant Garden', announced as a 'People and Plants' garden. It was explicitly designed as a space to rediscover the live knowledge of the forebears that 'influenced by Western medicine, was lost and forgotten' (figure 3.13). A hopeful future is alluded to in which a holistic healing process and a distinct South African culture that draws on all traditions might be developed. In this particular garden we can discover traditional healing plants, food plants, craft plants and magical plants. Solely

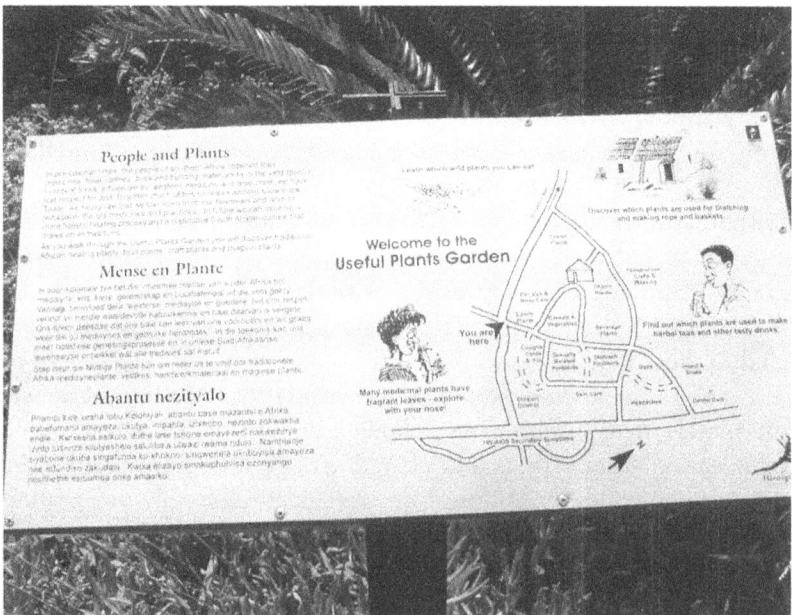

Figure 3.13. Kirstenbosch National Botanical Garden map of the 'Useful Plant Garden', Cape Town, December 2007

in this little specific area, designed for this purpose, are we allowed to smell, touch, taste and sense plants; however, we are left to do so unguided, or solely guided by bits of information we may try to mend together to make them meaningful.

Other attempts to make botanical gardens 'alive' are manifest in the Cape but have had limited success. A Rastafarian *bossiedoktor* selling herbs in her own shop in Cape Town told me how she partook in Table Mountain National Park's Medicinal Herb Garden project hiring 'local *izinyanga*' to keep the medicinal plant garden sustainable, letting the *izinyanga* use some of the plants for everyday use in return for tending to them. The woman explained to me how this experience had been a catastrophe for her, since they paid so little and for so few hours that they could not make the garden sustainable nor provide a livelihood for her. One of the problems was the distance between the garden and her home, as well as the strict 'visiting hours', another problem was the small fenced area in full sunlight. I visited other community gardens in the Rastafarian townships that seemed to 'work' to a certain extent, especially when the person tending to it lived nearby. As with laboratories, gardens can hence be more or less successfully inhabited.

Between the Delft laboratories high-tech machinery and the *bossiedoktor*'s backyard lie a variety of mediums in which humans engage with A. *afra*'s materialities. A. *afra* also travels through markets, shops, valleys, farms and festivals, again showing varying forms of engagements with humans.

Markets, Shops, Valleys, Farms and Festivals

Infinite quantities of dried plants are sold in bundles in herb stalls throughout the city. I walked through these markets every day, picking up pamphlets handed to me along the way. One Rastafarian *bossiedoktor* had a very small stall on the upper part of Cape Town's market, above the bus station, where I spent hours on more than ten occasions. Every morning he set up his materials, various herbs, always including small packets of A. *afra*, *buchu* leaf (Agathosma), wild garlic and, when available African potatoes and other seasonal herbs, roots and barks, along with kitchen towels and sometimes socks set out on a blanket with bright Rastafarian colours (figure 3.14). Mixing herbs with kitchen towels is telling of the commonality of wild plants for everyday use. All kinds of people came to him to

Figure 3.14. Rastafarian *bossiedoktor* medicinal plant stall in Cape Town market, February 2008

buy different plants, often sharing the ways they use *A. afra* for tea or snuffing and for which reasons they use it. This makes for a very lively knowledge of plants.

These markets hold everything from shoes to tomatoes. In one occasion the market was the location for an *isangoma* who I was accompanying to meet a client worried about jealousy within her business, offering her some counsel as well as some herbs. Mixing domains of the everyday in these public spaces is fascinating. The pamphlets that travel through these spaces promise to deal with love, work, AIDs or diabetes. In some you find healers naming their expertise: 'International Herbal Specialist on Aids Research', 'Spiritual Healers', 'Lady Doctor', 'Herbal consultant Dr. Mamba Ramathan here to wipe all your tears', 'Herbal Pharm & Consultants', 'Herbalist and Marriage Consultant'. Ghosts and demons in homes are addressed, court cases, high blood pressure, penis enlargement, sore throat and gambling. The latter reference to gambling strangely resembles our biomedical practices, which, through what some refer to as a process of medicalization, include more and more everyday practices as 'disease'. This seems to be the case with gambling, as well as with all issues of 'mental illness' and even drug addiction, as it progressively moves from being qualified as 'crime' to becoming 'disease'. Durban's medicinal plant market is considerably larger than Cape Town's and is organized in sections, including large areas of bark, but the market in Cape Town has more herbs. Mostly made of open temporary stalls set up every morning, there are permanent shops as well in the downtown area of Cape Town and throughout the townships (figures 3.15 and 3.16). In all cases, these plants are collected directly from the bush, dried and packaged in bundles for sale.

I visited numerous stalls throughout the city and townships, often to bring plants I had collected with Rastafarian *bossiedoktors*. Every couple of days I was invited to join in on a plant collection expedition with two Rastafarian *bossiedoktors* with whom I worked closely, sometimes accompanied by their families. We drove off in all directions to the peripheries of Cape Town in search of specific plants, roots and *Dassiepis*[3], and these excursions often turned in to full-day expeditions. Sometimes they had orders from *izangoma* who requested a specific plant from a particular area they knew or had envisioned. The locations were in the mountains, where access was still possible, and often in beautiful valleys that had not been developed for housing settlements or transformed through cultivation. The journeys to collect plants were very meditative, with long hours of carefully collecting the roots or plants selectively to make sure the plants could renew themselves (figure 3.17). With the Rastafarians,

Figure 3.15. Xhosa chemist's shop in Khayelitsha township, Cape Town, August 2010

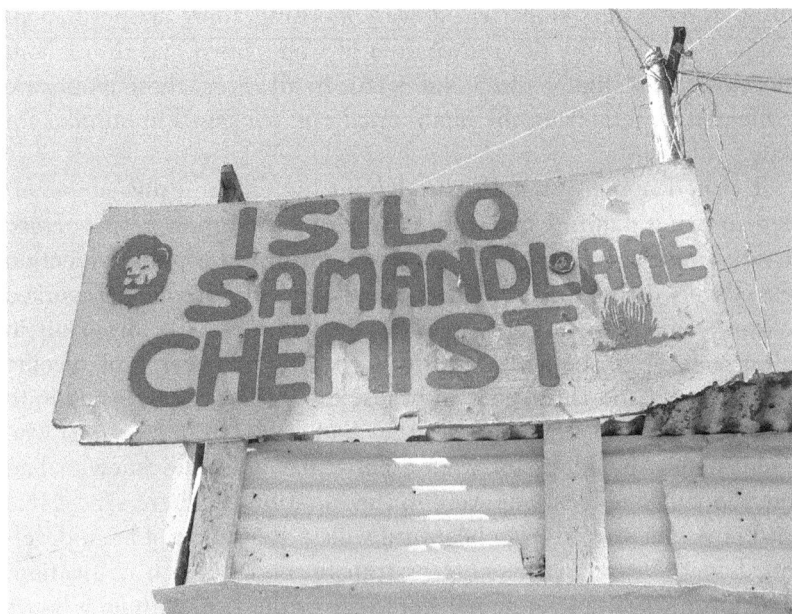

Figure 3.16. Xhosa chemist's shop in Khayelitsha township, Cape Town, August 2010

Figure 3.17. Gathering roots during a plant collection journey, outskirts of Cape Town, January 2008

I only ate raw foods during these expeditions, often wild plants and fruits we found along the way. The plants were often dried for a few hours before we packed them into the trunk of the car to store in a shed or bring directly to the *izangoma*. The Rastafarians were very romantic about the purity of 'nature', insisting on the beauty of the place, resting, sitting and discussing while the plants dried.

One day, after such an expedition, we stopped near Atlantis Dunes that had apparently been fenced off only recently (figure 3.18). All standing at the fence, we looked beyond towards the stunning dunes as the sun was setting. We could see a large tent set up a little farther on, apparently by a film crew. There was some discussion as to whether or not we should cross the fence. My son, who was nine at the time, stated we could not cross, since it was illegal. The Rastafarian woman we were with looked at me incredulously, stating that my son was quite a 'law-abiding citizen and that no one should control nature, which we are all part of'. While she was still saying this, my son slipped under the fence and began running through the dunes with his seven-year-old sister, the Rastafarian couple's three-year-old

Figure 3.18. Meditative discussions of relations with the environment, Atlantis dunes, outskirts of Cape Town, January 2008

daughter running behind them. We all laughed as we also saw some people from the film crew motioning to us that we should not be there. Our children had run to the top of a dune, and we thus had to go get them, entering the magnificent 'protected area'. It did not seem right however that the area was closed off to people, nor any other area upon which they depended for their livelihood.

Reid's apprenticeship with Rastafarian herbalists in Cape Town similarly tells how protected areas conflict with Rastafarians' work with medicine 'in the movement' (2014: 63). Not only are they viewed as a continuation of colonial occupation, yet 'Protected areas are founded on an epistemological foundation which views plants as static members of a population. In this framework, plant forms correlate with taxonomic identities which live in databases, journal articles, and books. Physical boundaries are mapped around plants to limit human contact and conserve species by maintaining their presence within the confines of the protected area' (ibid.: 62–63), very much a result of the kinds of research done upon plants as we have seen in the first chapter. Yet this limited access to plants is something they had to deal with on a daily basis, as new parks and protected areas emerged and with the increase in farmers fencing off their lands, sometimes specifically to stop Rastafarians from collecting wild plants on them. Everywhere the ratio of increased populations to decreased available land is a difficult issue to deal with. As plants become less available and less taken for granted, both ancient practices and new improvised ways of engaging with them however continue to emerge.

Rastafarian *bossiedoktors* are for instance very resilient. As I will discuss in further details in chapter 5, they not only occupy a centrally pacifying role in the townships of Cape town, yet they have become healers as well. Learning from 'the elders', explained as being the KhoiSan or 'first inhabitants', numerous have taken on the role of herbalist and work in collaboration with *izangoma*. Philander (2010) states there is an estimated two hundred bush doctors in the Cape, framing these practices as 'an emergent ethnomedicine' in her ethnobotanical study. While she immediately defends herself from the use of the term 'ethno' as an absolute, it is fair to state that this term nevertheless implies an underlying core or system. The Rastafarian movement has however, to the very contrary, appeared to me as a very open-ended way of being in the present. In its travels to South Africa, the Rastafarian movement has taken a place within apartheid as well as adjusted its search for roots in Africa from Ethiopia to the KhoiSan healing traditions and their knowledge in plants is fluid and lived rather than something to classify, fix or frame. The Rastafarian *Sakmanne* (bag men) going through meditative apprenticeship continuously tailor good relations with plants as a spiritual path, living almost entirely from wild plants, carrying plants with them to festivals, sitting amongst them as if they were part of them. One such festival organized by the Rastafarian community had whole areas filled with dried plants, roots and minerals as part of the event (figure 3.19).

Izangoma also entertain relations of proximity with plants, namely to connect with ancestors. There are numerous *izinyanga* shops in the townships of Cape Town, most *izangoma* also holding numerous plants, roots, barks, minerals and animal parts in their own homes or in nearby shacks. I mostly visited shops in the township of Khayelitsha with Rastafarian *bossiedoktors* either to obtain an order or to deliver plants we had collected for them. While I did not spend much

Figure 3.19. Sunsplash Rastafarian festival, filled with wild herbs sold by Rastafarian *sakmanne* (bag-men) spiritualist members of the Rastafarian community, Philippi township, Cape Town, November 2007

time in any of the shops, it was quite obvious they were part of the scenery, open to the street and highly frequented by the people in the neighbourhood. *Umhlonyane* was not particularly visible through the multiplicities of *muthi* in the shops, however it was often grown in the healer's backyards as well as collected in specific locations in the 'wild' for healing sessions.

While Xhosa *izangoma* and Rastafarian *bossiedoktors* keep an *A. afra* bush growing in their backyard for everyday medicinal use (see figure 3.12), they distrust bushes grown on farmland. Despite both being in the 'open air', there was a very fundamental incommensurability between healers and the farmer from Grassroots Group[4] growing *A. afra* for TICIPS's preclinical trial. It seems to have to do with ways of being in the medium. On the farm where *A. afra* is grown for the preclinical trial in the outskirts of Cape Town, the farmer took me to the field where the plant grew in ten rows in a small area of a large field where other plants were also grown (figure 3.20). *A. afra* was not identified for display, it was grown for mass production. He

Figure 3.20. *A. afra* fields, Grassroots Group, outskirts of Cape Town, December 2007

afterwards guided me to the separate locations where different steps of preparation of *A. afra* were conducted: the drying room where I found the plant laid down on boards in low-temperature ovens and the stocking room where *A. afra* was stashed in large bags ready for travel to the laboratories.

The farmer explained at length how he had agreed to grow *A. afra* for the laboratory studies, expecting healers to also buy some of his plants. To his surprise, the healers, however, never bought his plants, reasoning that the plants had lost their 'life' in this cultivated form. The healers seem to point to the fact that the farmer and his way of 'mastering' the plants by placing them in rows in a monoculture takes the 'living' out of them, very much as Merleau-Ponty critiques certain forms of science. I will return to this disjuncture between people throughout my discussions, since it indicates profoundly different ways of doing medicine. The way of the farmer is endorsed, even demanded, by the preclinical trial research consortium, which wishes to control and minimize the variability of *A. afra* as much as possible, while the ways of the healers rely upon this very variability, the liveliness, of plants in their mediums, which by all means include specific ways of engaging with them. These are the mediums through which I moved, following a particular trajectory that began with molecular biologists.

Quadruple Entry through Biochemistry

The mediums or worlds of people and plants I moved through are multiple, yet they connected through the particular trial that I began to follow in the fall of 2006. Searching for ways to enter the 'field' or a particular shared 'matter of concern' with regard to efficacy in medicine, I sent numerous emails out to scientists involved in any kind of clinical trial of indigenous medicine throughout Africa, a continent predetermined by my being a part of the Biomedicine in Africa research group (my 'field' was otherwise Brazilian Amazonia). This positioning included the criteria to orient research around an interest in how biomedicine was shaped, as well as how it was shaping this particular area of the world. When I sent an email to TICIPS', four biochemists part of the research consortium answered independently and in concert; they were respectively the directors of TICIPS, namely Quinton Johnson who was the South African director and William Folk who was the American director, the head of the Traditional Healers core of TICIPS, Nceba Gqaleni and the head of the Biodiversity and Phytochemicals core, Wilfred Mabusela. They all expressed shared concerns with varying notions of efficacy in dealing with healers and indigenous medicine in a quest to find the efficacy of *A. afra* against tuberculosis. This opening within the black box of efficacy during a preclinical phase, otherwise closed within the RCT procedure, was opportune; I hence joined the quest to find the efficacy of *A. afra,* or more precisely, to find what is going on in this process when indigenous medicine is part of the picture.

I met the first of the four 'gatekeepers', Wilfred Mabusela, in Germany in November 2006. He was the head of the Biodiversity and Phytochemicals core of TICIPS, professor at UWC and visiting scholar at the University of Potsdam when I was still stationed at the Max Planck Institute in Halle (Saale), a nearby city in Germany. The setting of this first discussion was in a desolate university building office near train tracks, a small room with a desk, two chairs and a computer. The biochemist went into detail about his work to determine the 'nature' of the plant, describing this task as a specialization in design, one of knowing the plant's constituents, compounds, active principles, biological therapeutic activity and chemical composition, as well as checking if the plant was consistent from area to area. He mentioned how selecting the molecular configuration to look into more deeply relied upon his experiences in the discipline, his intuition, as well as upon knowledge of the ways the plant was used and prepared by healers. Further, he spoke about the roles farmers played in nursing the *A. afra* plants cultivated for the trial. He explained how *A. afra* was initially grown on two farms, but had to be reduced to a single farm and farmer, in order to minimize variability and maximize the chances of finding a useful molecular configuration. He also, interestingly, led me to grasp some of the larger hopes surrounding his fundamental work.

Technically, his work was about finding the 'right' molecular configuration of *A. afra* to resolve tuberculosis within biological bodies; however, this was in light of a broader context. He then provided me with a first sense of the South African histories in which this trial was set, one that made the trial highly political, especially since it dealt with indigenous medicine. He elaborated on the sociopolitical aspects of delving into *muthi* through molecular knowledge as a way to gain legitimacy for the people who relied upon this knowledge, including himself. I was mesmerized by the sophisticated manner in which this molecular biologist described the trails of the preclinical trial as a political process, as a first step towards the recognition of the value of *muthi* by intermediaries linked with global health. Recognizing a paradigm where molecular biology holds the highest legitimacy within the RCT, he managed to tie his knowledge concerning the complexities of molecular effects to the increased possibilities 'to say and to act upon the actions of others' (Foucault, in Dreyfus and Rabinow 1984: 309). The trial was described as a gateway into new global networks as much as a quest for 'true' knowledge. I was absolutely enchanted by the depth of this scientist's knowledge of the sociopolitical meanderings in which his research was embedded. Already I could sense that science was closer to the ground in the South African context. It was also a much more sensitive ground and a very delicate place to walk through, since so much was at stake for some of the researchers involved. Scientific research had South African histories that, for some, meant the annihilation of indigenous medicine and, consequently, of indigenous people and traditions.

Engaging in a trial of an indigenous medicine was for him a way to retrieve the dignity of a people. While his work was at the micro-molecular level, his vision was macrosocial, and I was immediately hooked on following the trial in Cape Town.

While most events of the preclinical trial of *A. afra* were taking place in South Africa, following this trial also opened trails to follow in the United States, where the financing came from, as well as where numerous steps of the fundamental research were also undergone. TICIPS was part of a partnership between U.S. and South African scientists connected through a global health concern, yet the pandemic and indigenous knowledge were, and continue to be, in the Cape. The Traditional Healers core took care of most of the issues dealing with the collaboration with healers, in particular within the first research project TICIPS had undertaken in Durban (Kwazulu Natal, South Africa), done with another indigenous medicine, *Sutherlandia frutescens,* against HIV/AIDS. This trial, which Gibson (2011) describes as an ambiguous process and which I followed once in a while because it was part of TICIPS, added comparative value to my study, giving prospective value to what might eventually occur should *A. afra* undergo the RCT procedure; the trial on *S. frutescens* had completed Phase I (Johnson et al. 2007) and begun Phase II.[5] It implicates Zulu healers in different ways, which I will discuss later in this section. The Traditional Healers core of TICIPS aimed to coordinate these collaborations with the other TICIPS cores, which I will define in more detail below when I introduce my 'partner'. My own role was not defined in any precise way; however, it was tentatively guided by the second of the four 'gatekeepers', who warmly welcomed me upon arrival in South Africa.

The South African director of TICIPS and professor at UWC, Quinton Johnson, set the tone of my entry into the 'Cape Town part' of the field in a meeting during my preliminary fieldwork in February 2007. He largely led my way into the trial, at least to begin with. His wish to fit anthropology into the trial, by studying indigenous ways of making medicine 'work', opened a space for my presence. He is the one who partnered me with the head of TICIPS's Traditional Healers core. He also invited me to various meetings in both Cape Town and Durban during the length of my stay and in the following years, such as to the African Indigenous Medicine Systems conference in 2009. He helped organize my initial interviews with a plant systematist, a plant physiologist, an immunologist and a pharmacologist, all of whom participate in TICIPS's research consortium. Together we organized a series of seminars on herbal medicine with professors, namely with Diana Gibson, and students from both the Department of Anthropology and SAHSMI of UWC, which had been created in 2003 specifically, in his words, to 'unlock the value of indigenous medicine for wellness'.

During a meeting he organized, inviting his students from SAHSMI at UWC, which is the University known as the popular or 'coloured' University with a large population of Xhosa students, he seemed to have staged the way for one of his students to answer a question directed to me. He asked her, 'Why do we do research on medicinal plants if we already know they work?' His student answered, 'Because they [scientists] don't trust us.' It was clear that he already knew the indigenous medicine worked in healing and sought in the clinical process a route to legitimize what was already effective within broader circuits, more specifically within global health. His own relations with healers are in fact familial, his mother and grandmother both being healers. He was also a speaker at a ceremony at which certificates of participation were given to about 150 traditional healers who completed an HIV awareness course at UWC on 29 October 2009, showing manifest empathy with them. As with the MU-UWC partnership, the relations established through time within these 'educative/integrative' initiatives have the potential to eventually balance out the initial paternalistic teacher-student tone of such arrangements. It is clear, however, that who teaches who is a unilateral movement from the academic setting to the healers, with little movement the other way around. In other words, the healers' knowledge is not to be taught to the academicians; it is solely the healers who need to become acquainted with biomedical ways of healing, beginning with literacy.

Even the South African director's statement that laboratory studies were only done because of a lack of trust in indigenous knowledge did not appear to express a need to correct this obligatory route to follow; rather, he was satisfied to follow it and to guide healers along this path. This oddly did not appear to be a problem for the South African director of TICIPS, who, like the head of the Biodiversity and Phytochemicals core, also evoked becoming a molecular biologist as a path to gain legitimacy, as well as a way to move beyond the oppression experienced during apartheid. This lived oppression was mentioned in numerous subtle ways during our multiple encounters, as well as directly during a meeting on the waterfront terraces in Greenpoint in the centre of Cape Town, where at some point during our discussion he seemed to become silent and distant as well as change his tone of voice; he then mentioned how he was not allowed to be here as a child. Certainly he was aiming to make new kinds of histories, and the path he chose was to become a player, learning and conquering the rules of the game (in this case, the clinical process).

A wonderful businessman and philosopher, the South African director of TICIPS proudly showed the architectural plans for an expansion of SAHSMI. Our meetings in his office were often interrupted with visits by Chinese counterparts who brought new technologies or herbal remedies to his table, attesting to the expansion of his work with phytotherapies beyond *muthi*. Of utmost interest was his grand vision of the future of clinical trials. During our very first encounter, he drew current clinical trials of herbal remedies as a closed pipeline on a piece of paper: the conventional, standard RCT process, also named a pharmaceutical pipeline or drug pipeline, drawn as a completely airtight pipeline with 'light' or 'truth' only at the end of the tunnel, as illustrated on the cover page of the *Pharmaceutical Executive* magazine December 1st 2008 for instance.

On a second piece of paper, he drew the same pipeline with pores all along the pipeline showing the need for continuous input from social, economic, political and cultural realities in a new clinical trial design. The envisioned trial he drew was with multiple perforations, indicating the need for the trial to breathe through life during the process to maximize inputs from indigenous knowledge as well as to remain coherent with the context. Playing both fronts as a molecular biologist and a visionary man supporting indigenous people and ways of healing, he is very powerful. His vision of a clinical trial may well be one of the first steps towards renewing the RCT's standard model, and I will attend to this in detail in the following chapters.

The third 'gatekeeper' was another biochemist, William Folk from MU and the American director of TICIPS. I met him during one of his short visits in Cape Town in the fall of 2007 when I was settled there for my eight month fieldwork stay, as well as in his offices in the United States in 2008. His emphasis on the usefulness of the trial was on 'safety', even more so than on efficacy. While broader than the micromolecular aspects, his concerns contrasted with those of the South African biochemists in a fascinating way. His other leading concern after safety and efficacy was a romantic evocation of 'traditional healing' that gave flesh to the otherwise technical endeavour of representing the biochemical mechanisms of efficacy, which he did not even bother attempting to explain to me, perhaps since I was not from a discipline in the pure sciences. The few times I met him he spoke of an upcoming 'cow exchange' ceremony (*indilinga*) with Zulu healers involved in TICIPS's *S. frutescens* trial. In mentioning the 'cow exchange', his face lit up with some kind of spark, contrasting with is otherwise very serious, quasi-military tone of concern with safety.

He also manifested a fascination with what I might find in indigenous uses of *A. afra*, perhaps hoping I would delve into its intricacies to find ways of making 'cultural' features of *A. afra* useful for the preclinical trial. I did not live up to this expectation for a number of reasons, one being my own histories with issues of 'biopiracy' in Brazilian Amazonia. As is the case in India and South Africa, the histories of making medicine in Brazil attest to the very political, often heavily resisted, processes of making biopharmaceuticals; biopiracy is endlessly evoked as the theft of local resources by pharmaceutical companies, and I long ago took the decision to take this seriously, taking care to not simply collect and provide 'data' of a certain kind without envisioning the potential effects of doing so. Approaches such as ethnobotany and ethnopharmacology live up to these expectations, as they add what they generally refer to as 'symbolic' and 'cultural' information to the 'real' botanical, biological or biomolecular knowledge about the plants. This is precisely the kind of information that is found in botanical inventories and botanical gardens, making associations with plants, modes of consumption and disease entities.

As stated at length in chapter 1, my positioning is that the 'real' is not in need of being divided as such, especially not in regard to the recognition of indigenous medicine, and that this path is only a particular method followed by the RCT for which the preclinical trial is preparing. I am therefore not tempted to simply add 'cultural' variables to core biomolecular ones, nor can I extrapolate parts of indigenous medicine to feed into this 'objectifying' way of knowing. This explains how my assessment can seem to blur boundaries instead of adding 'variables', preparation modes or taxonomies to the otherwise technical or procedural methodology of the RCT. My objective is rather to answer more broadly to what is going on in indigenous ways of healing with medicine in a way that can be comparable to the way suggested by the RCT, or at least that can challenge and engage with it thoroughly, not superficially, or for it to disappear through the new filter of knowledge. In this approach I leave open possibilities for grasping other routes that are not necessarily predesigned but emerging, as well as possibilities for understanding how the current design of the RCT is brought into being (or not) with the specific preclinical trial of an indigenous medicine. For the U.S. director of TICIPS, the 'indigenous medicine' aspect of the trial seemed to provide a spark to an otherwise almost military concern with ensuring safety and efficacy through RCT procedures. The expectation that I could feed into this spark and bring more indigeneity into the rigid mold without threatening, challenging or questioning the very RCT research design was unfortunately not the route I intended to follow. At any rate, this route was already being undertaken by people much more knowledgeable than me, by, for instance, my TICIPS 'partner', the fourth 'gatekeeper'.

The head of the Traditional Healers core of TICIPS, Nceba Gqaleni, who I was partnered with, was deeply engaged in concrete discussions and negotiations with healers involved in the *S. frutescens* Phase II trial. This trial was being conducted in KwaZulu-Natal, where Zulu practices are heavily embedded in a majority of people's lives. I met him at the Nelson R. Mandela School of Medicine in Durban in December 2007, where he is a professor as well as director of the South African Research Chair in Indigenous Health Care Systems. Also a biochemist, he is nevertheless the speaker for the healers and described in detail the intensity of the ongoing debates. It was in the year 2000 that he first approached the Kwazulu Natal Traditional Healers' Council to encourage its collaboration in TICIPS's trial of *S. frutescens*. He explained the initial reaction of the healers was to say, 'Why now?': 'biomedical doctors have been working already for fifty years without taking into account the indigenous healers, biomedical doctors having very little knowledge about the reality of indigenous healing, even if all their patients also consult traditional healers' (personal communication, December 2007).

The magnitude, immediacy and novelty of a new virus threatening all people (HIV/AIDS), however, enabled the collaboration to begin. Healers and scientists have been involved in negotiations ever since. 'It is also the HIV emergency that brought King Mandu Chabala and public health together in 2003 in Durban', he explained (personal communication, December 2007). In October 2003, a Memorandum of Understanding was signed between the Nelson R. Mandela School of Medicine/University of KwaZulu-Natal (UKZN) and the KZN Traditional Healers' Council (including the Ethekwini (Durban) Traditional Healers' Council), Mwelela Kweliphesheya, and the Umgogodla Wesizwe Trust (Smart 2013).

The launch of the TICIPS trial in Durban began officially through an *indilinga* (roundtable discussion and cow exchange ceremony) performed between scientists and healers in February 2008. This significant ceremony involved the Zulu king Goodwill Zwelithini ka-Bhekuzulu, indicating that the healers gave their trust to the Nelson R. Mandela School of Medicine and were going to share their knowledge within TICIPS's research consortium. The healers were invited to discuss with scientists in an effort of collaboration. The passing of indigenous knowledge – when to harvest, what is expected from the plant, how the plant is prepared and used – rested on the *indilinga* permitting a sacrifice to acknowledge to the ancestors the passing of this knowledge, to acknowledge to the ancestors that times had changed and, finally, to ask for the ancestors' blessing for these new developments. A healer had explained to my partner that this ceremony was about bringing different ways of knowing together: 'It is like if you want to know about circumcision, you cannot know until you are circumcised. … These are different ways of accessing knowledge and both structural or theoretical organized thought and experience should be combined to maximize efficacy of a medicine' (personal communication, December 2007). It is a matter of building trust between different ways of accessing knowledge, and the *indilinga* was meant to enhance real epistemological discussions about the plant in question (*S. frutescens*) between TICIPS actors and Zulu healers.

The process of knowledge exchange, however, remains arduous, since the fortress of truth of biomedicine is worldly, and that of Zulu medicine is somewhat locally rooted and fragile rather than reinforced through global, national and legal instances. Different spheres of conflict are nevertheless explored around the HIV pandemic: how to reconcile the message of abstinence with a broader concept of marriage; how to reconcile the fact that healers do not have access to diagnostics within a diagnostically based infrastructure; how to deal with a virus that one party has no such understanding of. A large part of the process of dialogue between healers and scientists has been to train healers to learn how to write and document so that they can gain credibility by making indigenous knowledge graspable. It is the participation in this initiative that gave my fourth 'gate-keeper' the ability to lead the TICIPS Traditional Healers core.

As with the healers in Cape Town, the first big controversy had been about testing animals. Using mice and vervet monkeys in a trial made the healers remark that this was a waste of something precious, a useless step, since the indigenous knowledge of S. *frute-scens* had been done with humans, not on animals, especially not animals in a cage. Experimentation with animals crosses another ethical domain, in that every clan has an animal name and is responsible for protecting that animal, and these animals represent most of the local biodiversity. This disagreement, amongst others, had to be acknowledged yet also disregarded, since animal testing is a requirement to satisfy both clinical and governmental standards. Everything else in indigenous medicine that cannot be accommodated in the laboratory procedure, in other words, that cannot be measured, was also disregarded, leaving little else than a few hints from indigenous medicine to filter into the trial. It seems that healers mainly played the role of convincing their people to enrol in the trial as guinea pigs. In this sense it is the straightjacket of an ethics committee, the Food and Drug Administration (FDA), the World Health Organization (WHO) and the United States President's Emergency Plan for AIDS Relief (PEPFAR) that minimizes the possibilities of recognizing indigenous knowledge, because it does not fit the narrow RCT design and standard these organizations uphold, within others.

With just these four encounters, it seemed a picture beginning to emerge. It was clear how molecular biology was both a way to know a plant and to ascend to higher political orders. For the U.S. scientist, there was a paternalistic desire to protect and a romantic longing for exotic practices, even if seemingly humouring these rather than acknowledging their realities. All pointed towards complexities ahead in dealing with indigenous medicine in terms of issues of efficacy, yet also in terms of constraints found within the very clinical model that was predesigned, pulling towards desires for more grounded knowledge. A strong sense of justice also lay behind the whole endeavour, in particular from the South African scientists, who had all lived oppression through apartheid and had lived with plants and *izangoma* throughout most of their lives (with the exception of the head of the Traditional Healers core, who almost shamefully admitted his Christian upbringing and lack of knowledge about indigenous healing), pushing forwards the real hope of regaining human dignity for their people through the process. This however was not a straightforward process. The quadruple entry through biochemistry presented through encounters with the head of the Biodiversity and Phytochemicals core of TICIPS, the South African and U.S. directors of TICIPS, and the head of the Traditional Healers core of TICIPS provided solid diving boards from which to plunge into numerous trails and trials of *A. afra* in becoming a biopharmaceutical (or not).

Following Trails and Trials

Each of the four above 'gatekeepers' provided new trails and trials to follow. These progressively led me towards healers, first at the interstices of the preclinical trial and indigenous medicine, afterwards moving further away from the preclinical process and delving deeper into indigenous medicine. I will here introduce the key actors and telling moments, beginning with those experimenting with *A. afra* in controlled environments and ending with those practicing in-the-world. A middle part addresses what is negotiated (or not) at the interstices between these two 'worlds of becoming'.

In Thought

A plant physiologist at UWC, Alex Valentine, was pointed out as the first actor I should meet since, as the South African director of TICIPS explained, he is the closest to the plant 'in nature'. The plant physiologist explained his task as determining the 'nature of the plant'. The plant physiologist was involved in the comparative trial on *S. frutescens,* working towards understanding its efficacy on HIV/AIDS. His objective was to create the perfect environment, explaining that he was 'working to maximize the *Sutherlandia* active compound in a glasshouse'. The perfect environment is the one that maximizes the plant's production of one specific active compound, in this case the canavinine amino acid, a compound gained through a more simple chemical analysis of something that is already 'known' in the plant, as documented by another scientist. The plant physiologist had a student documenting the production of canavinine in the plant in 'nature', in this case as it grows in the 'Useful Plant Garden' section of the Kirstenbosch National Botanical Garden, discussed previously. Looking at the plant in the garden is a compromise given his inability to consider its extreme variety both in the 'wild' and in human use. He regretted two things: that he was limited to looking solely at canavinine, as he suspected other active compounds are also essential, and that he could not follow how the plant is manipulated in the 'open air' – how people harvest the plant, when they harvest it and what part they actually harvest for medicinal purposes. He strongly felt that much would be gained were he to acquire more knowledge about people's use of the plant in everyday life; however, as he explained, 'it is about selecting things that can be achieved in a time and financial frame'. They also must be achieved in the RCT frame.

Regarding the *A. afra* trial, I met a plant systematist (taxonomy) in the spring of 2008, namely Leszek Vincent, professor at MU and involved with TICIPS. He reported on the morphological aspects of *A. afra*. Being South African yet stationed in the United States, he explained how a lot of his work within the preclinical trial was to ensure a 'people glue' between the U.S. and South African scientists, a role he also played in some of my preliminary written works, on which he commented thoroughly as a peer reviewer. Otherwise, his work did not deal with the plant in 'nature'; he was given a 'standard' plant to examine solely in the laboratory. The infrastructure expertise in South Africa was found to be limited, and MU became the overseer of what took place with *A. afra*, with its more sophisticated laboratories and expertise. The plant systematist's laboratory had received a big batch of *A. afra* earlier in the year (120 kilograms), and he worked with this. This 'standard' plant came from Grassroots Group, the farm near Cape Town mentioned above. The founder of this corporation is a retired professor of agronomy at UCT, a bioprospector who gave up academics in search of, amongst other things, the best 'parental stock' of *Artemisia*. Ten *A. afra* plants of the same parental stock with maximal active compounds (containing at least 80 per cent of the volatile oils) were cloned and harvested in a large monoculture (see figure 3.20). Leaves and stems from this batch were used in the preclinical trial, and it was the biochemistry of plants from this batch that the biochemists, pharmacologists and immunologists studied. Ironically, and as mentioned previously, the Cape Town healers I met stated that these very plants were without efficacy, demonstrating a deep incommensurability with indigenous medicine from the very beginning of a foreseen clinical trial.

Another key TICIPS expert that I was referred to was an immunologist from UCT, Muazzam Jacob, testing *A. afra* on both mice and human cells infected with *M. tuberculosis* at the IIDMM laboratory of UCT, which is probably the most high-tech laboratory in Cape Town. With him I visited numerous laboratory rooms. One was vacuum-sealed. Another contained frozen blood samples of tuberculosis (TB) taken from South Africans diagnosed with the disease, which were to eventually be sent to the United States for more 'sophisticated' laboratory studies (or laboratories equipped with different machinery). He explained that

> TB frontline medicines are sixty years old, the newest thirty-three years old, and they need to be taken from six to twelve months, and for this reason, multiresistance strains of TB appear, as most stop treatments after three months. Time has shown the limits of the current pharmaceuticals, making the frontline drugs less effective. Vaccine trials are underway with the South African Vaccine Trial Initiative, also at the IIDMM of UCT, and numerous attempts to develop new drugs are ongoing to try to overcome this problem (personal communication, January 2008).

He sees plant medicines as another way to overcome the problem of resistance and perhaps as a way to find more sustainable healing practices. He first thought that looking for efficacy against TB with traditional medicines in South Africa was 'like looking for a needle in one hundred haystacks', but he now realizes that it is accumulated knowledge that has made its own funnel through time. The latter statement was brought into doubt by the plant systematist introduced above, who doubted that *A. afra* had passed the test of time, stating that 'evidence of efficacy are largely anecdotal; hence the need for the application of objective research' (ibid.). Immunology as a discipline seems to have a more fluid notion of life and knowing

than taxonomy, and this might help explain the contrasting positionings with regard to 'objective research'.

The apparent opening 'in life' in immunology is, however, accompanied by a subsequent closure, which appears in the immunologist's
explanations. I was at first enchanted when he began to explain
immunity[6] as 'something like life itself, being everywhere and nowhere', and afterwards became confused by his explanation that it
was 'something like a stealth bomber' (ibid.). While in the first case
I had imagined his fluid notion of immunity could provide a bridge
with the healers' notion of life found in plants, his metaphor of the
immune system as a stealth bomber simply seemed to upgrade the
military metaphor[7] with new sophisticated war technologies, that is,
following a notion of life as a movement of closure. He elaborated
that the immune system can circumvent the enemy, but cannot pro-
hibit it: 'it is more like a cloaking device' (ibid.). Bacteria would work
on the same basis as the immune system, as a detection system, a
radar system of the host. A drug is explained as the external intervention that tells the host to generate a response against the intruder
(virus, bacteria): 'sitting here we are being bombarded by bacteria.
We are not sick because we have intact immune systems (ibid.).'

He further explained that basic immunology attends to why a drug
works, if it works and how it works or can work to 'outwit' TB. To do
so, the effects of a specific dosage of an *A. afra* extract is observed
on the absorption of a very well-known and characterized virulent
live laboratory strain of *M. tuberculosis* inside an 'intact immune
system'. The 'intact immune system' is one that maintains a healthy
balance between the body and the environment. It is not the relation
between body and environment that will be observed, however, but
solely what goes on within the body, or more specifically, within the
cells in the body. Since it is in the cell that the effects are tested, it is
possible to assume that the 'intact immune system' is in fact located
somewhere. It is identified with certain organs and processes of the
body that provide resistance to infection and toxins, namely, the thy-
mus, bone marrow, and lymph nodes. The immune system as being
'everywhere' thus has closure within the human body, a body that requires defending and instructing with the correct information. The
only movement of opening that 'life itself' may have gained is fluidity
within the body as well as some strategic intelligence, yet one that
can be controlled, steered or reoriented through the proper stimuli.

The closed body excludes any possibility for the tailoring of bodily organization through skills acquired in the inhabited world.

Extracts of *A. afra* are used to see if they kill bacteria in test tubes; if they do, the plant is effective at this level. The extracts still need to be tried in a human cell that has been given *M. tuberculosis,* to see if they are absorbed in the cell (from the outside to the inside) without being broken down (in vitro). If the extracts can be absorbed, the next step is the preclinical evaluation of a host (more than the cell isolated from a body) with an intact immune system (the best model is a mouse, a guinea pig or a rabbit) that will develop the same symptoms (in vivo). 'In vivo' in this case is in the whole body (human or animal), as opposed to 'in vitro', in an isolated cell. 'In vivo' means in a living body, including the cells, which are themselves 'everywhere' in the body. 'In vivo' does not account for the broader context in which the bodies live and die other than as they might be represented in the cells (which is what is assumed). This is perhaps how one can refer to the immune system as being 'nowhere', since bodies are more than cells yet carry with them their histories of lived experiences. The complexity in moving from 'in vitro' to 'in vivo' lies in looking at a plant's effect in a living host that has its own 'intact immune system'.

The intact immune system is made to appear in the experiment as something stable and preexisting, implying that a specific drug may alter the organism in a specific way that is predictable and demonstrable through a method. In the event of severe disturbance, making one ill, or in the event of injecting a virulent live laboratory strain of *M. tuberculosis,* as done in the experiment, it is assumed that an external intervention with a drug will tell the host to generate a proper immune response. This intervention is strategic for the immunologist introduced above in more ways than in the experiment; it also signifies for him a great deal of ambition for a particular political cause. The demonstration of this mechanism is glued to a hope to retrieve indigenous dignity, even improve the national image. He compared the quest for a cure for tuberculosis with an indigenous medicine to the case of cardiologist Christiaan Barnard, who performed the world's first successful heart transplant in Cape Town, giving South Africa international pride; the South African wish to have its traditional medicine reach the global market was defined as a similar journey from South African roots to global uses. While pharmacologists, immunologists and molecular biologists agree that their number one objective is to find out if the 'medicine works' and if it is 'safe', this is regularly combined with an objective to recognize indigenous medicine through this same route. It is unfortunate, however, that this is achieved by mostly disregarding how indigenous medicine is done.

In parallel with the immunologist's work, a pharmacologist heading TICIPS's Clinical Trials core, James Syce from UWC, and whose study with James Mukinda is discussed in detail in chapter 1, is taking on yet another part of learning about the efficacy of A. *afra* by looking at the mechanism of action of one of the molecules of A. *afra* on a human cell. Pharmacology is the science of *pharmakon*, meaning 'poison' in classical Greek and 'drug' in modern Greek. It is explicitly designed to study the interactions between a living organism and chemicals. If the 'drug', whether man-made, 'natural' or an endogenous molecule (within the cell), has a medicinal effect, it is considered pharmaceutical. The pharmacologists involved in the preclinical trial of A. *afra* were the most mechanistic of all the scientists I encountered, relying heavily on the principle of causality. For them it was clear that people were ill physiologically, pathologically, and their genes, enzymes and molecules were the proper targets for altering, manipulating and modifying that status with the application of other molecules from the outside. What is best known and maintained as the only 'constant' within this equation is the designed molecule; 'the only constant that can be found is a regular dose of the medicine', as he stated (personal communication February 2008).

The head pharmacologist explained that he is only concerned with what happens pharmacologically between the identified molecule and the body. His concern is to make sure that the plant 'works' in the first place, clearly supposing it can do so on its own, notwithstanding human interpretation or manipulation, even if the way to know how it works is done through a man-made model and it is with reference to humans' reactions that we can define if it 'works'. Making sure the plant 'works' is initially an act of evaluating its lethal dosage, which is what was being done in the Mukinda and Syce (2007) safety study discussed at length in chapter 1. The toxicity of A. *afra* had already been tested on the vervet monkey native to Africa at the Delft laboratories (Muganga 2005), and it had also been pharmacologically tested on mice and shown to be nontoxic (Mukinda 2005). Once the established nontoxic dosage and form was established, then the pharmacologist would be concerned with how and why it works (or not).

The head pharmacologist was also very much concerned with the complexity and diversity of human reactions to medicines. Phase I of the clinical trial, should it occur, would be concerned with safety and security, what it does generally (proof of concept). One question that would need to be answered at this point is, 'Does the molecule do what it is meant to do?' (personal communication February 2008). This expresses how the experiment is about the verification of a hypothesis about how *A. afra* is already assumed to work. A second line of questioning is, 'Does it metabolize well inside the body?' (ibid.). Interestingly, this question supposes an active part of the body that needs to react accordingly to the way *A. afra* is supposed to work. It is, however, not the body-in-the-environment that is given agency; it is the physiological body itself, which is assumed to be universal. This is what needs to be demonstrated. The body or human cell is otherwise assumed to be passive, or awaiting stimuli. Mice, for instance, are assumed to provide significant homogenous results; their metabolism is assumed to be constant and to react uniformly to the same molecule in all biological bodies.

The very mechanical pharmacological approach is perhaps paradoxically closer to 'life in context' than that exhibited by the immunologists and biochemists, who were more visionary. The head pharmacologist explained that he oriented his experiments according to the way *A. afra* was used in the everyday. Beyond pharmacology, he also was involved in the ethics clinical trial committee, had served on the South African MRC board and looked carefully into what was feasible within the current RCT framework nationally. A master's project he supervised was to write a proposal to see if *A. afra* could go into the clinical trial phase and submit it to the ethics committee (van Wyk 2005). The proposal was refused, and the refusal was analyzed by a student, who showed how difficult it is to take medicinal plants to clinical trial inside the current RCT guidelines, which are designed for very specific pharmaceutical compounds rather than for whole plants as used in indigenous practices. This broader expertise brought the head pharmacologist to comment on the concrete difficulty of including healers within the preclinical trial, since they were not part of the experts recognized in the research proposal that obtained the grant to do research on *A. afra*: 'Healers did not write the grant proposal with us, so how can we involve them?' (ibid.). He commented further that 'other people were doing it [the clinical trial], why couldn't they [referring to his research team]?' (ibid.), thus expressing real pressures coming from indigenous politics in this undertaking in contemporary South Africa, as well as pressures stemming from following the RCT demands, and feeling caught between the two overlapping motions.

Interstices

Encounters between laboratory scientists and indigenous people in 'real life' are not always there, nor are they a prerequisite, even if working with indigenous medicine. Some of the actors involved in the peripheries of the preclinical trial, however, navigated through both routes. Gericke, the second editor of one of the botanicals referenced in chapter 1 (van Wyk and Gericke 2007), explained that he has had training in the *isangoma* school of knowledge, started a company of medicinal plants called Phytonova, worked for the South African MRC and acquired some authority on the subject of indigenous plants in South Africa. I asked how he reconciled *izangoma*'s ways of doing medicine with what is extrapolated in botanical listings. He felt no conflicts at this level, rather thinking that this was a necessary path. He further explained that healers ensured safety on a local level; however, RCTs were necessary to ensure safety on a population level, and documenting the plants was a first step towards this possibility. In this way *A. afra* could be used efficaciously by *izangoma* as well as documented so that it could be tested for a specific disease through an RCT. His preoccupation was with having more plants recognized through an RCT, in particular with relation to HIV/AIDS, and he lamented finding obstacles in regard to running RCTs and getting the treatments 'out there'. What remains a problem in this equation is that the RCT of an indigenous medicine overrules acknowledgement of healers' ways of working, potentially inhibiting their legal usage if shown to be inefficacious through clinical studies. Encounters between healers and scientists can have serious repercussions, often leaning towards benefits for the latter.

The first such encounter between laboratory scientists and indigenous people I participated in was a meeting organized in November

2007 by the manager of the Environmental Resource Management Department of the City of Cape Town to explain how the city managers were planning to organize the plant markets of the city and ask the healers their views on the topic. It was a classroom setting at the IKS branch of the MRC (see figure 3.1). The invited healers were selected based on the importance of their commercial activities. Two different groups of traditional health practitioners were represented: *izinyanga* (herbalists, strongly represented by Rastafarian *bossiedoktors*, although not exclusively) and *izangoma* (healer-diviners), which both use plants in their practices. More than twenty *izangoma* were present, as well as approximately ten Xhosa *izinyanga* and more than twenty Rastafarian *bossiedoktors*. The Rastafarian *bossiedoktors* were mainly present to fight for their recognition as traditional healers, namely as *izinyanga*. As of yet they do not have this official status, nor the privilege to obtain permits to collect plants in designated areas, as some healer associations have obtained. The healers expressed different interests and different particularities of their practices, distinguishing themselves from one another; yet all agreed that healing was not a business, and all were suspicious of the government's attempt to control their knowledge and their access to plants. Organizing the plant markets could eventually imply verifying the 'safety' and 'quality' of the medicinal plants, a process that brings in new experts overriding the healers' knowledge. The RCT is the very process that ensures this 'safety' and 'quality', and I have by now shown how even its preparation is not a neutral process, nor are these meetings, which appeared rather paternalistic.

The teacher-student approach from city experts/scientists/laboratory technicians to healers seemed unilateral in most of these encounters. This was manifest during this particular meeting when the healers were asked to fill out a questionnaire on the subject of their trading activities. This caused turmoil because of the format and inadequacy or complexity of the formulation of some of the questions. In this same approach I might recall the HIV awareness certificate provided by UWC as well as the workshops in traditional healing offered by the IKS branch (the latter perhaps being the most paradoxical, since it apparently involves novice healers training accomplished healers in their own trade, as I explained earlier in this chapter). At the Delft facility laboratories, the healers were similarly invited to learn the technical steps of plant transformation (see figures 3.7 and 3.8). They were also taught how they can/should have their herbal remedies tested by bringing them to the laboratories. They were invited to make a claim, and this would eventually provide them with some samples of the capsules, pills, tincture or powder of their plant in exchange for their knowledge. The healers were therefore invited to bring their plants to the centre, where the plants were then tested for their efficacy.

Upon undergoing this process, however, the healers often discovered things they were not expecting at all. Many were surprised, if not astonished, by the testing of a medicine on animals, since its uses are known with humans (notwithstanding the fact that the animals were in a cage, a context that exists nowhere), and some turned back at this point. Others were astonished by the need to test the plant for its efficacy since they already know how it 'works', as also mentioned by SAHSMI students, and then forsook the initiative altogether. Finally, what they eventually found out, and as written in very small font on the IKS website, is that they would also have to yield their intellectual property rights, their rights to their knowledge of the plants, their uses and their known efficacy in practice. The process of undergoing clinical trials is so costly that none would be able to finance such a process, which may last for five to seven years just to reach Phase II of a clinical trial, as well as cost a few million dollars (the complete RCT process totalizing up to seventeen years). This is a situation we are currently all confronted with, not solely the healers in South Africa; however, they might have more to lose in the process, since it can also result in finding their medicine toxic for lack of understanding in the way it is prepared or for looking only at a part of the plant and eventually inhibit its use. Were these huge financial investments instead be made to ensure access to wild plants, with deepened engagements with them, as done by a number of *izinyanga*, these encounters would be more interesting for the healers. Were they to be consulted for their expertise, this could be another interesting angle to explore. This is, however, not currently the case.

I met other groups of healers at an IKS meeting on the topic of trade and commercialization of traditional medicine. The opening statement of the meeting was that 'traditional medicine will be regulated one way or another, and we might as well be part of it'. What is implicit in this statement is that traditional medicine will be regulated through RCT procedures. The director of the IKS branch of the MRC, a businessman, an agent of Cape Nature (a nature conservation organization), a city manager, a natural medicine doctor, a Chinese doctor, three healers and I attended this meeting. Within the broader topics of infrastructure, access to resources, transportation regulation, recognition and control, and different areas of specialization in traditional healing, discussions about healers' 'secrets' arose: one of the healers was worried about releasing her secrets to others. Trust was once more involved: she was worried of being disrobed of her knowledge, which she did not reveal to anyone, let alone to laboratory studies. Since I had been introduced as the anthropologist 'without an agenda', I was invited to intervene. Spontaneously, I stated that scientific enquiry into her healing practices was actually incapable of robbing her 'secrets', since RCTs were not attending to relations established with plants, nor able to attend to the 'immeasurable' and thus to her very real ways of making the plants 'work'. They can, however, take very useful hints or data that can be very profitable as well as lead to regulations involving a plant. This meeting was held in January 2008, and it was already clear to me that because such a small amount of information regarding indigenous medicine, its knowledge and its uses could enter the RCT procedure, such secrets were 'safe'; however, the 'secrets' could be harmed through other regulatory routes.

A similar worry emerged with regard to being robbed not of 'secrets', but of the materialities of plants as well. This had to do with patents acquired on certain plants that were found to restrict their use in everyday practices. I was also asked to share my thoughts on this topic. Patents were in my view incorrectly understood as something that could inhibit the use of plants: patents are on a specific method of extraction of a specific molecule within a plant, usually inaccessible to anyone who does not work in a laboratory; thus, common usage of the whole plant, as done by indigenous people, was not contravening patent laws. This was not the way the agent from Cape Nature understood it, nor the way patent laws were politically 'publicized'. Patents are announced as a monopoly of a plant, even if technically this is not the case. On the other hand, RCTs are also announced as the pathway towards the recognition and legitimization of indigenous medicine, even if technically this is not the case as well. Indirectly or strategically, however, not only do RCTs not recognize indigenous medicine, they also have the potential to illegalize plants' usage should they be found toxic. Patents are similarly given monopoly over a plant that does not correspond to what is known or demonstrated scientifically. I return to these discussions in more depth in the following chapters, but first I want to share what I learned with healers in the mediums in which they worked, which for the moment are still outside these forms of regulation.

In Practice

My first visits with healers independently of meetings were with novices or their colleagues. I visited the young novice *isangoma/* undergraduate anthropologist who worked at the Delft laboratories facility offering the Traditional Healing workshops, in her home and healing room; the latter was a small, enclosed, extremely organized and clean meditative space filled with dried plants in glass pots, musical instruments, skins, minerals and other objects. I was told not to touch anything, but we could sit in the room to discuss at length her practice, how she connected with *muthi* and the spirits that spoke through her. A single mother, she also explained her challenges on an everyday basis of survival between work, school and child rearing. Later we set out with her colleague, a sociologist, also a healer apprentice working at the Delft laboratories facility in the Traditional Healing workshops, to meet with a young practicing *isangoma*. He also took me into his healing room to explain the rudiments of his practice. His working room had musical instruments, skins and benches for clients to wait on. Contrary to the previous healer, whose room could fit only two or three people, in this room there was space for ten to twenty people, on the benches and on the floor. In fact, he conducted both individual and group healing sessions on a regular basis. Interestingly enough, he was not particularly receptive to explaining his practice to me; however, he did so politely. I found that the accomplished elder healers were much more willing to share their ways of healing.

'I live with plants, they live in me' explained an elderly *isangoma* (personal communication January 2008). Her movements convincingly showed that there was no separation between her and her medicines, which were dried and kept on shelving along all four walls of her consultation room (figure 3.21). I met Mrs. Skaap in January 2008 during the meeting at the IKS branch of the MRC described previously, and she invited me to her home in the township

Figure 3.21. *Isangoma,* Khayelitsha township, Cape Town, January 2008

of Khayelitsha. She presented herself as a 'natural-born healer'. This was also stated on her business card, and she specified that she was from the Madlamini clan (the same clan as Thabo Mvuyelwa Mbeki, the president of South Africa at the time, 1999–2008), which links her with *ngoma,* or the larger practice of becoming a healer through a 'calling'. She was explicit about her knowledge of plants being incorporated into her own being, stating in contrast that 'those people who do research don't know much'. Her legitimacy was explained as follows: 'I am born and bred with plants, I live with plants. I received a calling, my mother, father, grandmothers are all *izangoma*' (ibid.) The eldest of her two daughters and her son are also *izangoma,* and her younger daughter serves clients when she is away. She also mentioned that she could not stand having a man in her life because it gets in the way of her work. Contrary to what is described as the 'core features of *ngoma*' by Janzen (1992) as well as by others (e.g., Wreford 2008), she told me she does not believe in ancestors, only in God the Creator (of the spirits of the wind, the mountains and water). She goes and prays in the forest and by the ocean. She showed me how the whistle spirits live in the upper-right corner of her consulting room and explained how 'they help cure and can also find and stop a person in a stolen car'. She further explained, 'I take the cockroaches, crocodiles, scorpions, needles … out of the body, sometimes with the help of the whistle spirits.'

Plants are used to move spirits around. *Umhlonyane* makes dreams clear. It can be used alone for the stomach or with *imphepho* (*Helichrysum sp.*) or pepper for cough. However, plants in themselves cannot be expected to cure. Guidance is needed to make them 'work'. She keeps the plants in containers, some fresh, most dried, and some in powder form bought in markets. She showed me stacks of written papers in files compiling her accumulated knowledge on different plants and (dis)ease; she hoped someone would someday type all the recipes down on printed paper to safeguard them. She explained health and disease as alive in the blood, which is why something needed to be extracted in the latter case: 'things alive, disease in the environment enter bodies and cause harm if weak' (ibid.).

What most surprised me during our discussions was her explanation that traditional medicines are so strong that they spit disease out, while pharmaceuticals can only soften disease: 'When someone is really weak he cannot take traditional medicine; he can go to clinic and then return to me to get cured, he needs to be strong' (ibid.). For this *isangoma*, contrary to what can otherwise be assumed because of their high chemical concentration, pharmaceuticals are weaker than wild medicines because their 'life', or powers, have been diminished. Ethnobotanist Wendy Applequist from the William Brown Center for Plant Genetic Resources at the Missouri Botanical Garden in St. Louis I conversed with, also linked with TICIPS, confirmed this in a discussion I had with her at the Missouri Botanical Garden. Using the example of aspirin, which is a single compound and more condensed, the ethnobotanist said that there is no proof that willow tea is less effective than aspirin, because the combination of multiple compounds in willow tea often makes it more efficient. She believed this should be taken into consideration for most plants, stating, for instance, that *Equinacea purpurea* (also from the Asteraceae, or daisy, family, like *A. afra*) roots are most likely more efficacious with earth on them than by themselves. These thoughts are anchored in practices that point in the same direction as the healers that claim that the *A. afra* cultivated for the preclinical trial has diminished 'life' as the plant moves from the bush to a controlled manipulation.

This can be demonstrated by the acknowledgements also made by molecular biologists regarding differentiated efficacies linked with place of growth, which is why the preclinical trial moved to utilizing plants grown on one farm by one farmer. We can also recall how the plant physiologist was disappointed that he had to only look at cana-vinine in *Sutherlandia frutescens,* because he was certain that other compounds linked with the place of growth and moments to collect the plant also played a role in increasing the efficacies of the plant. From another perspective, what Mrs. Skaap is also stating is that, to heal, the person also needs to be strong and thus be able to engage well with the medicine. In other words, for her, pharmaceuticals can attenuate the illness however it cannot heal in depth, which is what indigenous Xhosa medicine attends to.

The story of Mrs. Skaap is also telling of the high customization of *isangoma* practices, keeping some of the core features while bringing new ones into being. Healers deal with what is currently at hand, with what emerges during illness as well as healing. In a newspaper clipping of the *Big Issue,* a local journal Mrs. Skaap gave me during my visit (with no indication of the publication date), a story features her curing cases of AIDS as well as tuberculosis. The newspaper clipping also includes one of her quotes, stating, 'Sometimes my an-cestors tell me which herbs to use to help the dying soul', making her practice coherent again with core *ngoma* features, although this is perhaps not the main route she follows in healing. I am unsure if her negation of dealing with ancestors in our conversation had to do with a transformation of her practice or with her interlocutor (myself). She might also refer to ancestors in only certain instances, while preferring to evoke the whistle spirits in others. What is certain, however, is that her practice is customized and necessarily contin-uously changing according to what is lived around her and what is acceptable in her surroundings, including clinics. This makes her statements of utmost interest with regard to what is going on with medicine in her neighbourhood, in particular with regard to her thorough engagements in medicine, which seem to override every-thing else in the form of a certainty in her practice.

Another *isangoma,* Philip Kubukeli, a well known public figure in the township of Khayelitsha as well as in Cape Town and even in South Africa, a key *izangoma* in Wreford (2008), and also the healer I have been evoking in the previous chapters, expressed his connection with plants to me through their roots – the physical roots of the plant through which he connects with the earth, and also the ancestral roots through which he bonds with the known benefits of a medicine. He explained that a novice can only really work with indigenous medicines once he has found this crucial bond with his past and his surroundings. For *izangoma* this is acquired through intense drumming sessions in which the sounds reach the soul of the novice, 'releasing his inner strength' (Personal communication, February 2008). This was the essential point I needed to know.

With much less conviction than when referring to ancestral practices and with a sense of disillusionment, he had previously showed me some of his newspaper appearances, some of his certificates and his mostly disappointing attempts to join his practice with an official association in order to obtain the proper permits. He had joined the South African Traditional Health Practitioners, Herbalist and Spiritual Healers Association in 1994, an association not yet fully recognized; he is the founder and president of the Western Cape Traditional Healers Forum, president of the Western Cape Traditional Doctors, Herbalists and Spiritual Healers' Association.[8] Kubukeli was also, at some point part of various associations, such as the Healing Association of Southern Africa, the Traditional Chinese Healing Association, and the Hypnosis Society of South Africa, with only the South African Naturopathy Association still showing a trace per their website,[9] while he attested the existence of the other associations with old certificates pinned on his office wall. He had worked at Groote Schuur Hospital in Cape Town, at the Delft laboratories' garden transplanting plants, at the Kirstenbosch National Botanical Garden and in a botanical garden in Nyanga township. He also showed me the exams he made some of his students pass in their training to become healers (the pages I saw were the names of plants with a space for the trainees to inscribe the names of the disease that plant could cure, essentially a plant-disease association), but he also at this point stopped talking altogether. He looked at me and expressed physically that to know anything about any medicine or healing process, it all had to pass through the body. The rest seemed less important. The gesture he made with his arms to express how drum rhythms needed to be embodied to gain any knowledge about healing was most convincing. He immediately invited me to a drum session the following Sunday. I finally understood that he was telling me, in a number of ways, that papers and statuses had not much to do with his art of healing, and that knowledge and abilities in indigenous medicine instead had to do with the embodiment of skills and the ability to connect through mediums.

While I had spoken to *izangoma* during meetings and in their homes, visiting their working spaces, it was in assisting this drumming session that I began to understand how *isangoma* practices 'worked'. When I arrived at the drumming session, accompanied by a Rastafarian *bossiedoktor,* lots of things were already going on. People were moving about in all directions, preparing food and placing objects. New people were arriving every couple of minutes. Through the commotion, I was welcomed to sit in the living room and to meet with some members of the family. After an hour or so, the *isangoma* disappeared in his bedroom to change his attire. When he returned he glanced at the video camera I held in my hand and spontaneously stated, 'So be it', signifying I could film the event if I desired. A few minutes later we were invited into an adjacent room, where I was told to sit on a long bench together with the other forty or so people attending. An open space was left in the middle. The ceremony began with two women setting the rhythms on the drums. Different women entered the circle at different moments, dancing and talking in tongues. The *isangoma* appeared some twenty minutes into the ceremony, also singing, dancing and talking, with only a few words in English manifestly addressed to me, at thirty-seven minutes into the song, stating: 'I am a literate man, but when I accepted the grandfather's spirits …', and after this ten-second acknowledgement of my presence, he returned to Xhosa words and song.

After more than an hour, the initiate entered the room. She was a woman in her late forties who was passing through what Janzen names the 'white', here explained as the first of seven steps towards becoming an accomplished *isangoma*. A pot was placed in the middle of the room containing *ubulawu* (dream foam). The *ubulawu* was mixed with a stick, and it was publically confirmed that we were connected to the ancestors and all was going fine. A family member stood near the initiate, seemingly approving the passing of his relative into the *isangoma* practice. The singing, drumming and dancing, as well as various gestures involving white necklaces placed around the head of the initiate, continued for more than two hours, with varying crescendos and calmer moments. The initiate had been sit-

ting and kneeling in the middle of the séance during the whole process, only standing when a new gesture was accomplished, such as
placing a necklace. After three hours people slowly began to fizzle
out of the room and move into the kitchen and living room. I was
invited to drink beer in the living room, and asked to contribute
some money for the event. Someone had been sent to provide *dagga*
(cannabis) for the Rastafarian *bossiedoktor* who was accompanying
me (knowing he did not consume alcohol), and we were offered food
to share with them.

The drumming ceremony was the *isangoma*'s answer to my question, 'How does *umhlonyane* work?' The plant was nowhere to be
found, and perhaps this was the point; indigenous healing was not
about an object-plant. It was later explained to me that *A. afra* had
been set under the bedsheets of the initiate the night before to prepare her for the ceremony. It served as a way to 'purify' the initiate
so that she would be able to connect properly with the ancestors the
following day and begin her journey to become a healer. Although the
isangoma had shown me a list of usages of *A. afra* for coughs, colds,
headaches, diabetes and tuberculosis, the role of the plant was also
more subtle, less visible, a medium for connecting with the world.

Kubukeli has been a practising *isangoma* for almost fifty years. He
was part of the team of traditional healers who participated in the
drafting of the Traditional Health Practitioners Bill, signed into law
in 2007. In a newspaper article he states, 'Things won't change all
of a sudden. This is a big industry. What will happen is that finally
we will get some recognition and upgrade our standard of healing'
(Ndaki 2005). After following the hopeful route of the institutionalization of traditional medicine for more than twenty years and becoming literate in the process, when I met him in 2008, he had taken
the decision to return to his roots, or to his grandfather's spirits, as
he states – a stance Merleau-Ponty (1945: x) would qualify as 'preobjective', as it embraces nonstandardized states of consciousness
and indeterminacies of bodies in the world. During the drumming
session, Kubukeli had stated his literacy in opposition to an older
form of knowing anchored in skills, and explicitly acquired through
the lived experience of the body. Literacy, on the other hand, is associated with science and representational or propositional forms of
knowing, such as in the papers and exams he had shown me. It was,
to him, a diversion from the knowledge he had acquired through

the body and ancestors. This seemingly necessary distinction of objective and preobjective is at the heart of the preclinical trial of an indigenous medicine, because it leads directly into the bodily politics currently lived in a South African context with the advent of humanitarian biomedicine and its histories.

Propinquity

Proximity with plants and embodied knowledge is key to the workings of indigenous healing. In encounters with both *izangoma* and Rastafarian *bossiedoktors*, it was the relations they had with my husband and his musical abilities that really struck me. My husband was immediately recognized for his skills in knowing through embodied sounds, while as an academic it was much more arduous for them to try to answer my requests, or even recognize my ways of knowing or wanting to know. I even stopped enquiring directly about *umhlonyane* in these contexts because my quest(ions) seemed to be missing the point entirely. *Umhlonyane* was not an object nor was it to be objectified; it was a medium to engage with and through which to intermingle as part of the world. This required a new approach. To simply ask about *umhlonyane* was to suppose it was something 'out there' to know from a distant, disbanded standpoint, since it externalized it. It was less about asking about *umhlonyane* than getting to know it through bodily cultivation. I also began to follow other *muthi* and how people engaged with them, how these were collected, prepared, presented and consumed in homes, festivals, city markets and during mountain treks, as well as in laboratories. I needed to be immersed in the complexities of the world if I was to grasp any understanding from *umhlonyane* in indigenous medicine. This also explains why I had been invited to share in the world of the *izangoma,* to enter into their 'particular synaesthetics'. To me, this was an essential part of the way to learn how to heal with plants; it was part of the healing process and also the most grounded way to understand how *umhlonyane* 'works'.

Assisting a drum session was transformative. It caused me to seriously rethink how ways of conducting research might affect human relations. It took me into phenomenological literature to find more engaging ways of doing research. In other words, I felt that my intrusive enquiry was inappropriate, misplaced and inadequate. It had to be done otherwise. While it did not seem to hamper relations with scientist colleagues, and indeed seemed to fit those relations (since they share similar traditions), approaching *isangoma* healing practices intellectually, or by maintaining a certain distance, not only did not fit, but also hampered relations and possibilities of learning anything from indigenous medicine. This transformation made me move inwards to be best able to reenter the world in new, more intermingling, ways. While this was not entirely new to my own research, as I had applied reflexive approaches in the past, it here required me to rethink how research more generally was affecting socialities, as well as orienting knowledge in particular ways. This undertaking made it all too obvious how the process of withdrawal from the world, which is required to prepare a plant for an RCT, is such a unique route.

This is how it became clear to me that a whole array of ways of engaging with the world was being lost in the RCT process. Not only does the RCT disregard these important practices of making medicine work, it explicitly demands these practices be left behind. In the quest for knowledge about *A. afra,* scientists are led to create innovative man-made medicine through a fabricated model. This model implies the repetition of precise acts in order to eliminate all but one relation; that occurring between the designated plant molecule and the human infected cell. This one relation supposes universality brought through distance and the reduction of the complexities of the multiple practices that come together to make medicines work (or not). *Izangoma,* for their part, seem best at steering complex combinations of relations between people, plants and environments in the past, present and future. It is through proximity, immediacy and connection that complexity is dealt with 'in life', and it is in this way (dis)ease is reoriented through embodied skills. Multiple mediums become ways of healing roots of (dis)ease as well as routes towards new relations within the world with others. In preparing towards following the RCT model, scientists need distance, objects and mechanisms to represent the efficacy and safety of a molecule that only exists in laboratory conditions. Despite these standardizations, in both cases embodied histories play a role, and in both cases preestablished standards play a role. Yet the role of standards and embodied histories is almost in reversed order: embodied histories are central for the *izangoma,* while they are something to filter out as much as possible for the scientists; preestablished standards are at the heart of scientific enactments, while the work of *izangoma* is to move away from them as much as possible in order to realign the relationships involved. For the indigenous knowledge linked with A. *afra* to be understood thoroughly in the preclinical trial, it would need to include sounds, smells and gestures as well as account for enhanced embodied skills, performativity, reflexivities and sentience, which is acquired not through the accumulation of mental information, but through intensified bodily engagements in the medium and the world.

The preclinical trial assembled things and people with which I moved as well. I followed what the trial announced it would do (provide a route to recognize indigenous medicine), and ended up finding what escapes it – that is, most of the indigenous, hands-on and embodied knowledge of what makes A. *afra* 'work'. How it works, and lives, thus became much more important to me than A. *afra* 'itself'. Through these mediums and in looking for a route to 'really' recognize indigenous medicine, my research is less about inventory, collecting data, describing acts and providing representations (which could inform the RCT procedure only superficially) than about acquiring the skills favourable to understanding how medicine is kept alive, is activated, used and made to be useful. Life is kept in plants through proximity as well as interacting with the 'wild', as opposed to controlling it; 'wild plants' can mean 'plants out of control of humans', as Ingold underlines with regard to 'wild animals' (2000: 62). In this case, the notion of 'wildness' is rather one of propinquinty, of kinship. In answer to issues of overharvesting, deforestation or the closure of forests for 'protection' and 'conservation', usually excluding humans, an *isangoma* explained that live knowledge in plants actually ensures efficacy and minimizes uncertainty through proximity and a deepened awareness of medicine in life.

While a scientist needs a laboratory as a medium to make medicine work, healers need grounds and people to work with in their mediums; healers are ultimately concerned with different forms of life as movements of opening. Both solutions are worthy to bring into conversation. Most importantly, this is the hopeful route that I have proposed to follow. I have begun to do so by describing the mediums in which people engage with medicine and what it feels like to move in them. I then introduced the ways I entered the preclinical trial as well as the trails it led me to follow, in particular with regards to the 'indigenous medicine' it aims to recognize. I demonstrated that the preclinical trial is unable to attend to the bodily skills required to understand indigenous medicine, constraints manifested by the scientists simultaneously led to follow the RCT model requirements. I intend to pursue this conversation by exploring further which kinds of life are dealt with in indigenous knowledge, even in simply imagining indigeneity.

Notes

1. The UM, United States and the UWC, South Africa, have more than seventeen years of bilateral exchanges through the UM-South African Education Program, a historic program that provides the foundation for TICIPS. See http://ticips.missouri.edu/about/facts.html (accessed 25 August 2014).
2. The other core sections are Administration (led by two biochemists codirecting the other cores); Clinical Trials; Communications; and Biodiversity and Phytochemicals. See http://ticips.missouri.edu/about/people.html (accessed 25 August 2014).
3. *Hyraceum* or *Dassiepis* (in Afrikaans) is the dried concretation of hyrax (also called dassies, they are a small, herbivorous mammal) urine, which has a long history of medicinal use by the Khoi (van Wyk and Gericke 2007: 131). The Rastafarian *bossiedoktors* I worked with kept some in their garage and stalls. They mostly sold it for fertility purposes, telling me about the case of a couple who could not have children and had come to them all the way from Europe to try *dassiepis* and it was a successful therapy.
4. Grassroots Group (Pty) Ltd works on Grenvley Farm in Gouda, Western Cape Province in South Africa, approximately 40 minutes from Cape Town. From modest beginnings in 1992, Grassroots Group became a market leader in the use of South African plants. See http://www.grassroots group.co.za/index.php?id=121 (accessed 26 August 2014).
5. See note 1 in chapter 2 for details on the RCT phases.
6. It is immunity upon which *A. afra* is found to have effects in the Ntutela et al. (2009) study, slowing down the progression of tuberculosis through antimycobacterial activity.
7. In the science of immunology of the 1960s, the military metaphor became a notion linked with the successes of the smallpox vaccination and today remains the dominant metaphor for what vaccines do: combat disease. According to Haraway (1988), the military metaphor, imagining vaccination as the dispatching of an army to fight the enemy inside the body, a metaphor that is still perpetuated in public discourse today, is a political strategy of the administration of populations camouflaged in the biological (biopolitics). From earlier definitions explaining immunity or the immune system as a defence mechanism protecting the body against foreign agents,

some fluidity and agency seems to have been allocated to the living cells. Haraway (1988) found, within the immunological discourse itself, notions of the body that link the somatic to the biological. The specific flexibility of the immune system entertains a notion of the body in interaction with the world and for which body the borders of disease are ambiguous and indefinite (Laplante and Bruneau 2003: 526). From the early to the late twentieth century, biomedical bodies have shifted from organism to biotic component (code), vaccines from magic bullets to immunomodulation, biological determinism to systems constraints, and concerns in microbiology and tuberculosis turning towards immunology and AIDS. These are just to name only a few of the historical shifts in the production of bodies in biomedicine, as noted by Haraway (1993: 376).

8. See http://www.zoominfo.com/p/Philip-Kubukeli/2178894 (accessed 26 August 2014).
9. Per their website: http://naturopathy.org.za/ (accessed 26 August 2014).

Chapter 4
Imagining Indigeneity

In essence, 'tradition' was defined as the absence of biomedical tools, titles, packaging and substances – the antithesis of
what was white and biomedical.

– Karen Flint, 'Indian–African Encounters'

Clinical trials of *muthi,* or broadly speaking of any South African
indigenous medicine unrecognizable in biomedicine, are undertaken
with the hope of retrieving the dignity of a people through recognition of their knowledge. RCTs give hope of a way to finally prove
to the rest of the world that a 'traditional' medicine works. Adams's
(2002) research in clinical trials of indigenous medicine in Tibet
made her call the process one of randomized controlled crime. In
her research the trials of an herbal remedy can only benefit those
who set the rules of the game. For the Tibetan doctor, only two negative outcomes are possible: either the medicine is judged inefficient
and its use can fall into the legal arena and potentially become illegal; or the medicine, or one or two unrepresentative active principles
of a much more complex combination of plants, is judged efficient
and will be transformed into a pharmaceutical product owned by a
multinational corporation, with little local benefits (Adams 2002:
669–70). She further explains what happens during the process of
an RCT that makes the game so uneven. First of all, the disease the
medicine is tested for is not necessarily the same entity as the one
defined by the healer – diagnostic categories not being universal.
Second, it is assumed that a well-designed RCT produces reliable
evidence free of interpretation; the healers would have most likely
had another analysis of the results at many steps of the process.
Third, RCTs assume that treatments can be reduced to discrete lists
of isolable active ingredients, as most traditional medicines are com-

binations of plants. Finally, RCTs rely on the placebo approach, or on the exclusion of the placebo, whereas some claim (e.g., Waldram 2000: 617) that by expelling the placebo from the controlled trial, one of the main pieces of evidence for the efficacy of traditional medicine is therefore ruled inadmissible.

Renewed interest from international agencies have made the instances of clinical trials of 'indigenous medicine' occur with more frequency on a worldwide scale, reviving old debates and challenges as to how to define and provide access to the healing benefits of medicines as health commodities. The 'objectivity' of RCTs becomes more difficult to make and uphold in attempts to prove the efficacy of 'indigenous medicine'. The required processes of standardization, measurement, demonstration and representation found in an RCT are challenged by deepened engagements in the world as they play themselves out in the everyday or in particular staged performances. The trial of an herbal remedy in which healers are involved is an instance at the very nexus of varying healing strategies.

In this chapter, I begin to explore what it could look like to attend to indigenous medicine beyond botanical inventories and what implications this may have with regard to clinical studies. I first attend to the broader histories of 'indigenous medicine' as an ontological category in anthropology. I follow this with histories of how traditional medicine was included within the World Health Organization (WHO) and, more recently, in specialized institutional organizations that are directly involved in the preclinical trial under study, namely, the National Center for Complementary and Alternative Medicine (NCCAM) of the National Institutes of Health (NIH) in Washington, D.C., and the South African Indigenous Knowledge Systems (IKS) branch of the South African Medical Research Council (MRC). Afterwards, I delve more specifically into *muthi* histories and presents, more particularly evoking how Xhosa *izangoma* (healer-diviners) and Rastafari *bossiedoktors* (*izinyanga* or herbalists) work and collaborate in keeping knowledge between people and plants alive. Some specifics of what occurs when *muthi* is prepared for clinical trials are then discussed, with concluding remarks on how to avoid standardization in *muthi*. What is formally recognized in *muthi* from the RCT point of view is solely a part of a plant, bark or other part of 'nature' that can be made into an object, a bioresource or a commodity. From another perspective, *muthi* is broader than its 'objects', which are not

closed and reified nor reducible to healing commodities. Biomedicine is also broader than its 'objects'; however, it relies heavily on
processes of universal objectification to broaden its scope. My aim
is to eventually bring these practices into the conversation, as hoped
for and partly done in the preclinical trial of an indigenous medicine
discussed in this book.

Indigenous Medicine

The terms 'indigenous', 'indigenous medicine' or 'indigenous knowledge' are not without their histories of troublesome connotations
and resulting inclusions through politics of exclusion, a phenomenon that Lash explains as constitutive rules pertaining to the second modernity, whose key institutions govern exclusion (2003: 5).
Traditional medicine (TM), indigenous knowledge (IK), indigenous
knowledge systems (IKS) and, more recently, complementary and
alternative medicine (CAM) are all terms used more or less inter-
changeably to refer to all forms of medicine that are not scientific,
or, more precisely, that are not supported by 'scientific proof', which
means they are not defined and demonstrated through experimentation in an RCT. These forms of medicine are acknowledged, however, in a special category, which makes some scholars remark that
'perhaps the most damaging use of the term "indigenous knowledge"
is that it sets IK in opposition to "scientific knowledge"' (Green 2008:
134). To avoid replicating this divide, and as explained in the first
chapter, I address knowing as skills of engagements in-the-world,
attending to what both making 'scientific medicine' and 'indigenous
medicine' may 'sound' and 'feel like' in practice, since this is how I
was able to grasp how it 'works'. I further make a distinction between
this form of knowing and information acquired about plant parts and
constituents and their effects on human physiological processes,
thus somewhat taking the empirical back into life-making processes.
In this way, I aim to bring 'indigenous medicine' into conversation
with biomedical ways of making medicine 'work', not as exotica or
requiring translation through the RCT filter, but as contemporary
practices that challenge and feed into current ways of knowing in
science and research. In line with what a number of scientists involved in the trial indicated, attuning attention to knowing in practice invokes a new kind of trial, perhaps not an external trial at all.

Finally, the term 'indigenous medicine' has taken on a life of its own, which also merits attention.

Anthropologists have played and still play an important part in the 'coming into being' of the notion of 'indigenous medicine'. I myself brought the notion into being where it did not formerly exist in some of the indigenous Amazonian villages where I conducted previous research (Laplante 2004), already asking myself at that time if it were necessarily beneficial to create this opposition. Numerous anthropologists and missionaries had done so before me in other locations perhaps, with similar preoccupations about making 'indigenous medicine' visible in spheres where tending to health and well-being was otherwise generally assumed to not exist at all, let alone in a coherent manner. Early classical ethnographic descriptions of 'primitive' societies have informed how healing can work in other contexts. Rivers (1924) provided a pioneering account spreading the idea of coherence in primitive ethnomedical practices, a tradition that continues in various subfields such as ethnomedicine (Nichter 1992), ethnobotany (Balée 1994; Schultes and von Reis 1995) and ethnopharmacology (Beaucage et al. 1997; Etkin 1990, 1993; P. Roy 1996); the latter two subfields have been mostly preoccupied with 'natural' biochemical properties of the effects of plants within their perceived respective taxonomies.

Early involvements of anthropologists in international health programs (1945–60) were more concerned with promoting biomedicine (Cora DuBois first held this position with the WHO in 1950), decrypting cultural codes to facilitate the introduction of biomedical ones. Anthropological research was at that time mainly situated in what Foster later called the 'adversary model' (in Coreil 1990: 9), a model systematically opposing traditional thought to that of the modern, aiming to understand the habitus of the first to accommodate a replacement by the second. This approach was smoothed down through the image of the 'empty vessel', which offered health programs a simple, educative project in contexts assumed to be without any health care because of their lack of biomedical technologies and services; a cognitive approach that is still today largely advocated and applied in public health programs abroad. The period between 1960 and 1975 saw a decline in the involvement of anthropologists in international health programs, only to see it reemerge with more intensity following the WHO's 1978 endorsement of the need to recognize 'traditional medicine' (Laplante 2004: 41). The question of the recognition of traditional medicine then became central to the definition of the role of the anthropologist, a role that was taken seriously in the emerging field of medical anthropology, yet perhaps not in the direction predicted.

According to Good (1994: 27), the first anthropology explicitly 'medical' was not formulated to encourage the efforts of public health interventions, yet as a critique of their cultural naivety. Paul's (1955) work would be fundamental in the definition of medical anthropology in those terms. An even more critical wave emerged in the 1960s, with politico-economical critiques leading this movement, which persists today. The 1970s saw the beginning of introspection in biomedicine, but it would take until the 1980s for an important movement of medical anthropology to address the culturally embedded practices of biomedicine. The works of Leslie (1976) and Kleinman et al. (1976), for instance, demonstrated that biomedicine is one system within others, making the hegemonic pretentions of biomedicine more difficult to uphold. Coexistence of diverse medical systems became a centre of interest, Leslie (1980) being the first to introduce a notion of 'medical pluralism'. At the same time, Last (1981) proposed thinking in broader terms of 'medical culture' to grasp a particular hierarchy of both systematized and less systematized forms of knowledge in relation to medicine as they appeared throughout his work in Africa. Last's account was echoed a decade later by Pool (1994), who made a plea for the dissolution of the concept of 'ethnomedical systems' and, more recently, by Littlewood (2007), who understandably argued that the time is overdue for medical anthropology to question the concept of a medical system that underlies much of the work in medical anthropology, namely, finding the notion of ethnomedical system to be mostly in ethnographers' minds. I argue this is also the case in assuming a biomedical system. Rather, I find open-ended 'knowing-in-the-making' in both indigenous and biomedical healing practices, something that is always emergent and done or undone by both human and nonhuman actors entangling in-the-world.

Perhaps the most influential approaches in medical anthropology were those that turned directly towards biomedical practices as a way to understand 'indigenous medicine'. The pioneering ethnography in a laboratory done by Latour and Woolgar (1988), investigating approaches in science and technology studies, facilitated this turn towards biomedical practices. A proliferating Foucault following simultaneously began to be concerned with biopolitics as a way to grasp biomedical practices as well as find ways to move beyond them (Rabinow 1996; Fassin 2000, 2006a; Lock and Nguyen 2010), failing, however, to attend to 'indigenous knowledge' in the process. Foucault (1976) describes modern forms of governmentalities or biopolitics as an increased investment in knowledge about bodies through the disciplines (in biology, for instance) enabling to make decisions about ways to administrate and to regulate populations. For numerous scholars, this marks an era of 'biopower', one that would have expanded its scope through increased 'development'. The latter industry came to rely on discourses and institutions that crystallized during 1945–60, a period during which the relations between key variables such as capital, technology and certain institutions (universities, governmental) forged together (Escobar 1984–85).

These discourses on development metamorphosed into discourses and institutions of humanitarian medicine by the end of the twentieth century (Hours 1998), largely facilitating the expansion of biomedical practices across the world. The place occupied by traditional medicine in these networks oscillates, sometimes given more credibility and space, other times dismissed to various degrees, according to various strategies. This means that the borders between traditional and cosmopolitan medicine fluctuates. In the case of their encounter during the clinical trial of an indigenous medicine, these boundaries are to be redefined, indicating potential areas of relevance in reshaping views of evidence in global health. This is of fundamental importance with regard to how the clinical trial of an indigenous medicine may take place, as well as how anthropologists may translate this knowledge (or not), either to fit within a scientific model, as a distinct 'ethnoscience', or as a way of learning about healing-in-life, as I propose.

From Evans-Pritchard's *Witchcraft, Oracles and Magic Among the Azande* (1937) to Lévi-Strauss's *La pensée sauvage* (1962), healing routes have been described as respectively unique and universal, yet nevertheless recognizable. The 'primitive' has long been found informative of simpler beginnings of the modern world, yet mostly perceived as relics of the past, even if contemporary. Much is, however, learned and exchanged through these fieldwork experiences. Mead

(1985), for instance, showed how living with the Balinese made it feel possible to breast-feed her child for a longer period upon her return in the United States. Favret-Saada (1977) became an apprentice sorcerer to hear words in witchcraft, necessarily transforming acceptable truths in her everyday. I partook in numerous healing rituals with *izangoma*. I used numerous plants, fumigations and teas to heal a variety of cough, flu and wound issues I or my children suffered during fieldwork (Laplante 2004) and upon our return home. I also shared a few things I learned along the way. I have discussed in-depth how these transformative routes took root in medical anthropology's turn towards embodiment in chapter 2. These worthy routes of shared learning indicate commensurability in the everyday, while other routes are more arduous. The RCT model, and those preparing and following it diligently, sets itself on a position of 'verification' of the efficacy of indigenous medicine, but on its own terms. Anthropologists have tried to understand indigenous medicine to demonstrate how it might work within other terms, with the continuous risk of setting 'indigenous medicine' into a category of its own in contrast with 'real' medicine, a feat that was reenacted within the preclinical project discussed here to specifically test and recognize 'indigenous medicine' within a scientific tradition. The struggle with this dilemma is contemporary and is, perhaps, beyond 'indigenous medicine' and 'biomedicine', as it has more to do with the ways 'knowing' is interpreted.

Ellen and Harris describe indigenous environmental knowledge through ten characteristics (2000: 4); the first four points refer to indigenous knowledge as being context-dependent, rooted in a particular place and set of experiences, and generated by people living in those spaces. Agrawal argues, however, that 'attempts to draw a strict line between scientific and indigenous knowledge on the basis of method, epistemology, context-dependence, or content … are ultimately untenable' (Agrawal 1995)(2002: 293). For example, 'on the point of its context-dependence – or the idea that IK is relevant only to a specific context – Agrawal points out that the sociology of scientific knowledge demonstrates the extent to which scientific truths are context-dependent' (Green 2008: 134; see also Agrawal 1995). In the preclinical trial's offshoring from the United States to South Africa, it appears that scientific knowledge is bounded to global networks of legitimacy, making it context-dependent and also rooted in a particular place (reproduced as best as possible in varying locations) and a set of experiences. Bridges can be established with regard to methods and approaches in science that are closer to indigenous knowledge than others, making the strict divide not between indigenous medicine and science, but between varying theoretical approaches and corresponding practices; for instance, I aim to show how a phenomenological approach in anthropology is very close to numerous indigenous healing practices. Science is also done within the realm of the lived and the felt; however, it generally attends to these ways of knowing in a more or less tailored fashion, most often with an intention to disengage from these. As such, the approach I propose enables one to find a common ground to attend to the ways 'indigenous medicine' and 'biomedicine' are brought into being rather than as preexisting systems dictating practices. Understanding practices as continuously emerging diminishes rather than increases the opposition between indigenous medicine and biomedicine. This is an uneasy path, since the category of traditional medicine, as set in opposition to science, seems to be strengthening rather than weakening.

Wahlberg (2006) notes a 'revival' of traditional medicine in the past three decades accompanied by a growth of interest in the issue in sociology and anthropology. He finds three predominant anthropological and sociological approaches accounting for the history of traditional medicine, all of which emphasize traditional medicine–biomedicine dichotomies to varying degrees (ibid.: 126): a first 'is often rooted in a classic critique of modernity (not least its medicine) as life-enfeebling, alienating and dehumanizing which is duly contrasted with the vitalizing, emancipatory and rehumanizing potential of TM and CAM' (ibid.); a second series of approaches he notes tends 'to account for the history of TM and CAM in relation to bio-medicine in terms of a politics of (self-)interests between rival groups, movements or professions'; 'finally, a third common form of distinction between TM/CAM and biomedicine in sociological and anthropological studies centres on the question of their legitimacy, which in turn is dependent on concepts of "efficacy"' (ibid.: 127). The latter approach often cites a kind of Kuhnian epistemological incommensurability (ibid.: 128). Walhberg's solution for forsaking the traditional medicine–biomedicine dichotomies found in these approaches is to study it, in his case the recent history of traditional

herbal medicine in Vietnam, as a field of *problematization*: 'not from
the point of view of politics, but always to ask politics what it has to
say about the problems with which it was confronted ... [to] question it about the positions it takes and the reasons it gives for this'
(Foucault 1997: 115). Following what is going on in the preclinical
trial of *A. afra* is a similar enterprise; however, I take it a step further
with Foucault's predecessor, Merleau-Ponty.

To a certain extent I am interested in understanding what Wahlberg formulates as 'the unavoidably normative grounds that under-
pin the ongoing elaboration of "safe", "effective" and "proper" ways
of using traditional medicine' (2006: 128); while this normative
ground is supported by a predesigned model and often distant agencies such as ethical committees, clinical trial monitors and federal
drug agents, I suppose that all of these actors may or may not make
the same norms appear in the same ways since practices are continuously emerging in new ways. In a similar attempt to avoid the
dichotomy between indigenous knowledge and science, I aim to do
so by adopting a phenomenological stance, grounding knowledge in
practices that precede both domains. This is how I propose to move
beyond the indigenous knowledge–science dichotomy, which I argue
ultimately rests upon a culture-nature dichotomy, usually placing
'indigenous knowledge' in the 'culture' category and 'science' as simply reflecting 'nature'. In this way, and as also stated in the previous
chapter, I do not aim to 'add' more culture or humanity to biology or
to the RCT, nor do I aim to romanticize indigenous medicine to the
detriment of biomedicine, as in the first approach mentioned above
by Wahlberg, but I aim to understand how medicine is being done
in life-making processes in both cases and how these routes might
inform each other more profoundly. I also do not take the issue of
traditional medicine and biomedicine as a politics of self-interest
between rival groups as a point of entry, yet I end up there in following the politics of knowledge that are brought into being in the
preclinical trial. Finally, while I am interested in issues of legitimacy
and did enter the preclinical trial through epistemological concerns
that arose with the issue of 'efficacy', I move beyond epistemology to
address issues of ontology, or how ontologies are done or undone in
practice, and it is phenomenology that brings me there. In this way
I first assess some of the histories of attempts to standardize what is
made to appear as complementary and alternative medicine (CAM)

in the United States and as Indigenous Knowledge Systems (IKS) in South Africa, showing how 'indigenous medicine' plays itself out through these national and transnational meshworks.

CAM versus IKS

The preclinical trial of *A. afra* was a project initiated and financed by the NCCAM branch of the NIH in Washington, D.C. With its interest in African phytotherapies, the preclinical profile and terminology easily slips from CAM, usually of 'industrialized countries', into a fascination with TM or IK, usually of 'developing countries'. As mentioned earlier, the notion of TM appeared in official documents for the first time in 1978 in the Alma-Ata Declaration written in the context the WHO International Conference on Primary Health Care. In this declaration, 'traditional medicine' appears as a basic human right and as part of the 'health for all' objective, broadly defined as 'ancient health practices linked to a culture which existed before the application of science to health questions in opposition to modern official scientific medicine or allopathy' (Bannerman et al. 1983: 9). In this definition one can read the spirit of the endeavour to 'save' lost ways, as well as the view that science replaces ancient ways. In the WHO's 2008 definition,

> traditional medicine is the sum total of knowledge, skills and practices based on the theories, beliefs and experiences indigenous to different cultures that are used to maintain health, as well as to prevent, diagnose, improve or treat physical and mental illnesses. Traditional medicine that has been adopted by other populations (outside its indigenous culture) is often termed alternative or complementary medicine. (WHO 2008a)

To be fair in the use of this definition, biomedicine would then be 'alternative' or 'complementary' in its travelling to South Africa, for instance, a country where other kinds of medicine prevail. This is, however, not how things are played out. While from the 1983 definition to the 2008 definition we see 'knowledge' appear as opposed to solely 'culture' with regards to 'indigenous experiences', 'culture' still does not seem to apply to 'scientific medicine', which belongs to the domain of 'nature' and is thus implicitly assumed to not be culturally embedded. It is very difficult for a dialogue with IK to occur in these circumstances, circumstances that are reinforced through a sturdy model. Again, as with the 'placebo', which is known to have therapeutic benefits, indigenous medicine is also known to be beneficial, however in both cases they are to be set aside to leave the space to 'true knowledge', evidence or proof of real efficacy. As argued earlier, it would take even more than simply acknowledging that science is culturally embedded to modify this current organization; this would require thinking outside the very dichotomy between nature, made by science, and culture, to be made natural. In the meantime, institutions dealing with TM are still organized around this opposition. Legitimacy of 'true' knowledge is exclusively reserved for biomedical ways of healing, which remain the reference against which to test the efficacy of all other forms of medicine in global health networks and in this way overrule any 'real' recognition of other acceptable rationales or ways of engaging in medicine.

The NCCAM in the United States and the IKS in South Africa differ in that the first appeals to 'alternative and complementary' ways of healing, while the second refers to 'indigenous knowledge', meaning practices entangled in everyday practices of the greater part of the population. Bringing the two attempts to standardize 'traditional medicine' into conversation in this section is useful, since the preclinical trial of *A. afra* is an initiative funded by the NCCAM, which clashes with South African efforts to bring 'indigenous knowledge' to the forefront not as an alternative or complementary ways of healing, but as knowledge that needs to be given dignity in its own right through its ancestral roots. While the NCCAM initiative is less threatening to the status quo of RCTs, mostly reproducing them as best they can, and is simply a way to make more medicines legitimate and available through national and transnational health organizations, the IKS initiative is potentially much more challenging to the RCT standard, yet perhaps also more enriching. To 'know' and acknowledge indigenous medicine within the RCT model is a step further than to simply disregard its existence; however, acknowledging indigenous medicine with a project to 'test' its veracity without utilizing the means or learning the skills to grasp how it may 'work' keeps the incentive stationary or, at best, allows only a hint of indigenous medicine to be included into the wider scientific web. Once the 'hint' is included in the preclinical phase, everything else related to indigenous medicine appears as negative or as placebo, if it appears at all.

While this way of doing things can be a problem in itself, since minimal feedback from the world can fit such a model, it becomes a greater problem when it involves indigenous knowledge that is alive and well, such as is the case in South Africa. South Africa's 350,000 traditional healers are consulted by 80 per cent of the population (Flint 2008: 183), making these practices more than worthy to acknowledge, although not to overrule using a single model. 'TICIPS recognizes the value of this indigenous knowledge, the pressing need to ensure that it is preserved, and the importance of recognizing the contribution of indigenous knowledge and equitably sharing benefits derived from its use' (TICIPS 2005). In practice, however, the preclinical trial of *A. afra* recognizes very little, since the scientific practices undertaken are generally taken as a given, while indigenous practices are either well-known but unable to filter through the model, acknowledged superficially as exotica in a politics of exclusion or simply unattended. These three positions play themselves out in the preclinical trial of *A. afra,* but it is only the last that will transpire in the later clinical trial phases.

The NCCAM appeared as a branch of the NIH in the United States in 1998. Since then, its status has remained fragile and relatively weak within the NIH. I was warned not to mention too many of the issues it was confronted with in testing indigenous medicine. Although it was never explained clearly how my research could threaten the establishment, I suspect the NCCAM is under pressure to apply the RCT guidelines even more strictly than in testing synthesized molecules without acclaimed roots or histories, since it might otherwise threaten the very model. Already acknowledging the existence of such therapies is contested terrain. To name traditional medicines 'alternative' and 'complementary' is a partial solution making them less threatening to 'true' knowledge or the status quo, somewhat including them through their exclusion. The struggles of translating mind-body therapies into 'measurable objective evidence' are enormous. Therapies found to work in practice need to be translated into a new 'experimental tradition', which may hinder the possibility to understand their efficacies. While processes undertaken to test the efficacy of CAM do propose new models, they never overcome the ontological divide between nature and culture, which they leave entirely intact and even perpetuate.

Verhoef et al. (2002), for instance, proposed to add qualitative re-search methods to the 'gold standard', in this way only adding flavour to primary qualities rather than addressing the lived and the felt. In another article, Verhoef et al. (2005) introduced mixed-methods research with an explicit aim to reach equal esteem for qualitative as well as quantitative approaches, attesting to the current hierar-chy between these methods in research practices, which clearly have come to rely on the latter. A study by Vuckovic (2002) shows how pain was measured through both qualitative and quantitative studies, with the final results established on a scale of one to ten. While the evaluation of pain may be done in a more integrated way, its ultimate result, provided by a number from one to ten, is perhaps even more harmful than simply disregarding lived and felt experience, which is lost and distorted in the outcome. Ultimately, the legitimacy of CAM will be acknowledged through dissociation from its contexts via an external positioning rather than deepening relations in its contexts, as I will later propose in line with how indigenous healers proceed in healing. The tug and pull between reaping some of the benefits of being legitimate on a worldly scale or being relieved of some of the burdens and rationales of medical regulation is even more intense in the South African context.

In South Africa, the recognition of indigenous knowledge is played out at the national level, with all the problems and benefits that come with it. Its greatest feat is to become visible and malleable, leading to potential conversations within the political climate as well as win-ning certain legal freedoms, in particular with regard to land claims and in defence of environmental knowledge, as argued by Chapin (2004: 21). The biggest problem, however, is that indigenous knowl-edge has been set in opposition to scientific knowledge in a poli-tics of exclusion as well as closure. Indigenous knowledge has today taken shape as a political struggle, largely instrumentalizing essen-tialist notions of culture: 'the mobilisation of the idea of "indigenous knowledge" is itself a vehicle of modernity, and a political instrument in the era of globalisation' (Green 2008: 133). In this line of argu-mentation, the RCT model does not discriminate in its procedures, giving to all knowing-in-life the same discredit, including physicians' and scientists' experiences in healing and in making medicine for healing. The problem arises when the RCT is announced as a way to recognize these practices, such as in the preclinical trial I follow in this book, which in this case appeals to human dignity in-the-world. The RCT model is clearly not a pathway to do so, although it creates an occasion to revise its very model.

The South African Indigenous Knowledge Systems Policy originated in 2004. The National Indigenous Knowledge Systems Office has recently announced a plan to bolster indigenous knowledge systems in a similar fashion as was done with the NCCAM: the plan is to design an academic degree in IKS and to establish a pilot centre at the University of Zululand to be used for recording, codification and dissemination of IKS. The acquisition of this information, similar to the kind of information found in botanical inventories as discussed in the first chapter, will facilitate running eventual clinical trials of traditional medicines. Numerous trials of traditional medicines have already been run by the IKS branch of the MRC, sometimes mixing South African, Chinese and Brazilian combinations of the 'top stars' (respectively, *Sutherlandia frutescens*, *Panax ginseng* and *Paullinia cupana* (guarana), for instance); the director of the IKS branch of the MRC currently holds a Patent Cooperation Treaty and a South African patent on seven novel compounds isolated from a medicinal plant with antimalarial action. Clinical trials of traditional medicines at the national level generally follow international guidelines. In practice, the IKS branch of the MRC only runs clinical trials until Phase II, stating that this is sufficient for the benefits of the nation, and that Phases III and IV solely benefit global health networks (Matsabisa 2008). The IKS branch offers workshops on traditional healing given by healers, or rather, apprentice healers, as an accomplished Xhosa *isangoma* explained to me regretfully. The IKS branch also invites healers to bring to them claims of the healing properties of some remedies. The IKS branch of the MRC aims to be competitive globally, 'going back to our roots for innovative health solutions' (Javu 2011: 1). The redress of health traditions will rest on the WHO's four categories of the most important issues affecting the practice of traditional medicine: safety; efficacy and quality; access; and rational use, which I have shown were not neutral processes. TICIPS's trial also follows the models supported by these organizations, leaning towards a similar unilateral form of knowledge translation as proposed by the NCCAM, the NIH and the WHO.

The South African Traditional Health Practitioners Bill, signed into law in 2007, moves in a similar direction in its aim to provide a regulatory framework to ensure the efficacy, safety and quality of traditional health care services. It similarly takes for granted that evaluating efficacy, safety and quality is a neutral process. The bill promotes a traditional health practice that complies with universally accepted health care norms and values and aspires to improve the quality of life of the general public. These regulatory initiatives show a clear path from tradition towards biomedical and scientific standards, which is the route also followed by the preclinical trial I follow. How adopting this path is a way to promote indigenous knowledge remains unclear, since the process of legitimizing a medicine through an RCT is precisely one done by disregarding roots and histories, even its own.

TICIPS's project of running RCTs on indigenous medicine can be further found to fit with former president Thabo Mbeki's 'African Renaissance' project as a strategic political gesture to gain legitimacy in a global health culture. The African Renaissance is for some 'a part of the politics of image (hence of power) of the regime of Pretoria which, described as a model of "non-racialism" and of democratisation, makes itself the eulogism of the renaissance of the continent' (Crouzel 2000: 171). The African Renaissance is largely based on South African politics. It was evoked by Nelson Mandela a few times, and the idea was formulated by Vusi Mavimbela (1998) in a document entitled 'The African Renaissance: A Workable Dream.' Thabo Mbeki afterwards adopted this idea of a 'renaissance' as a way to reinforce his legitimacy within the African National Congress, a 'traditionalist' political party of the Republic of South Africa. The African Renaissance can be understood as a pan-Africanist variant of the twenty-first-century desire to confront the challenges of globalization in an international order (Kornegay and Landsberg 1998). 'It can be seen as both a culture and a doctrine of emerging international politics' (Landsberg and Hlophe 1999). Ferguson (2006: 114–15) sees it as an ephemeral moment restricted to the moment of its emergence (1997–99). The African Renaissance nevertheless refers to a renewal of Africa in terms of its strategies of 'Africanization' and of communication largely founded on a desire to reestablish links with African roots.

In his capacity as the chairman of the Presidential Council on Traditional Medicine, Herbert Vilakazi formulated the strategic political gesture as follows in an address to former president Thabo Mbeki:

'Such herbal mixtures, from African traditional healers, would, if verified, bring about a mighty revolution in modern Western medical science, and would bring honour, reverence, respect, and pride to African people and to the African continent. This would be a mighty force within the concept and program of the African Renaissance and of the New Partnership for Africa's Development (NEPAD)' (Vilakazi 2006). The Indigenous Knowledge Systems Policy of 2004 explicitly situates indigenous knowledge as the 'affirmation of African cultural values in the face of globalization' (Indigenous Knowledge Systems Policy 2004: Introduction, sec. 1.1), a dynamic that is at play in TICIPS's preclinical trial I follow in this book, as it is hoped it will promote African dignity as well as reach global networks; the RCT is perceived as the lawful process to reach both of these desires. Earlier South African histories, however, might help understand how the RCT has difficulty upholding its 'objective' status as well as its promise of African dignity.

Muthi Histories

Umuthi, which means 'tree' or 'bark' in the Bantu Zulu language, and it also means 'medicine'. The word is rendered as *muthi* or *muti* throughout South Africa, loosely applied to all forms of medicine that are not obviously biomedical, pointing in particular to Zulu and Xhosa *isangoma* healing practices, yet currently often narrowed down to the more recognizable and recognized *inyanga* (herbalist) practices. *Izinyanga* are also generally recognized as coming from Zulu and Xhosa traditions, but *inyanga* practices are also largely performed by herbalists of Indian descent along the Indian Ocean coast and in the Cape, a role also visibly undertaken by a large number of Rastafarian *bossiedoktors* who root their practices with herbs in both the Rastafarian movement born in Jamaica and in KhoiSan ancestral traditions in Africa. *Muthi* includes bark, shrubs, plants or herbs, as well as animal fats, skins, bones and minerals and/or chemicals, even seawater; all of these materialities are open-ended entities with multiple potentialities. The same medicine can, for instance, both produce and cure the same symptoms (Bryant [1909] 1966: 57–58). Beyond being sold as commodities in markets across town, *muthi* may be activated by spirits, applied across spatial distances and do not necessarily have to enter the body (Ashforth 2000).

Throughout her descriptions of 'being called' and her journey becoming an *isangoma,* Wreford refers to her *muthi* as *ubulawu* herbs (2008: 129). *Ubulawu* is a combination of plants and bark mixed into water that creates a dream foam, which has different roles within the process of becoming an *isangoma,* including for cleansing purposes and to enhance communication with the ancestors. *Ubulawu* can be used in different ways. In Wreford's case, 'the *ubulawu* acted both as emetic and as a wash to cleanse the external body. Each *thwasa* (initiate having received a calling) learns the herbal mix for her own *ubulawu,* how to grind and grate the roots and barks into her tins, how to beat the liquid into its frothing head, and, in Nosibele's evocative phrase [her mentor], "how to eat it". The exact combination of herbs remains a secret, not to be divulged to anyone' (ibid.: 110–11). Thus, a specific mixture of *muthi* is prepared specifically for the healer and with relation to her ancestors; the names of the *muthi* are only to be known as the graduation of the *isangoma* draws near. For these reasons, which indicate the deepened engagements between the *isangoma, muthi* and ancestors, I contend that *muthi* is indissociable from the acts, performances, divinations, spells, songs, dances, ancestors, skills, abilities and knowing that make these materialities potentially beneficial as mediums to weave good relations-in-the-world in healing. *Muthi* is, however, often narrowed down to the plants sold by *inyangas.*

It is thus interesting to note that the term *inyanga* had a much broader meaning in the early nineteenth century. Flint's historical analysis of white observers' notes from this period shows how *inyanga* meant more generally 'doctor' or 'specialist', usually single-remedy doctors;

> an *inyanga yezilonda* (*elonda*: sore) was thus a doctor who specialized in healing sores; an *inyanga yonzimba-mubi* (*umzimba-mubi*: bad body) – an abscess-doctor.... [T]he more general term *inyanga yokwelapho* (*elapha*: to treat medicinally) was recorded by Dohne as early as 1857 and defined as a master of administrating herbs. Another term for a general herbalist was *inyanga yemithi* (*mithi*: plural for medicine), a healer who treated bodily ailments medicinally or surgically. (Flint 2008: 52)

These last two meanings seem to be those that are used in contemporary South Africa, excluding surgical practices. What is also interesting in these early accounts is that healer-diviners, today called *izangoma,* were also called *izinyanga,* namely, *izinyanga* who worked on the national level. Examples include an '*inyanga yokumisa izwe,* or doctor for "making the land stand firm", who treated the national *inkatha* (symbolic grass coil) which secured the strength of the nation. An *inyanga yezinsizwe* (*zwe:* nation) or *umsutu* strengthened the army and nation through medicine' (ibid.: 52–53). As with the term *inyanga,* which has come to denote the general herbalist, in the mid- and late nineteenth century so too the term *isangoma* came to refer to any type of healer-diviner. The term *isangoma* is of Zulu origin; its Xhosa equivalent is *igqirha* or *amagqirha.* The term *isangoma* is nevertheless the title generally used throughout South Africa and Cape Town, even by the Xhosa healers encountered during my study. *Isangoma* was the term used by the Xhosa healers to refer to themselves (both diviners working with ancestors and with whistle spirits), while the Rastafarian *bossiedoktors* or herbalists were claiming the status of *inyanga,* mainly for permit purposes to access lands to collect plants. 'The categorization of *inyangas* and *isangomas/isanuses* or the conjoining term "traditional healers" adopted in the twentieth century is a manifestation of African's colonial experience, which only further intensified encounters with healers from various areas' (ibid.: 66). Of interest is how the two domains have remained separate, the latter remaining more difficult to standardize as well as legalize within biomedical rationalities. While all types of healers were criminalized by white legislators in the 1860s, only midwives and *izinyanga* were licensed in 1891, singling out *isangoma* practices as the most threatening.

Biomedicine's struggle for legitimacy in South Africa was explicit. It was highly reliant on legal measures and politics of exclusion of traditional healers to make a place for white medicines and practitioners (Flint 2001, 2006; see also Shapiro 1987). The precolonial period of the Zulu nation (1820–79) already marks interconnections between African and European therapeutic practices: 'though the Zulu kingdom remained politically independent until its defeat by the British in 1879, it was not unadulterated by white influences' (Flint 2008: 38). White traders and missionaries were present throughout that period and, in 1838, a third white community – the Boers – settled along the western edge of the Zulu kingdom. According to Flint, 'the decline of the Zulu kingdom (1820–79) and the rise of urbanization, migrant labour, and consumer culture resulted in the disappearance of specialized knowledge of herbs, gathering techniques, and medical practices' (ibid.: 39). How new influences are indigenized (or not) can explain, for example, affinities with inoculation on the one hand and refusals of techniques such as amputation or even pills on the other hand (ibid.); the smallpox epidemic in 1863 was, for instance, dealt with by the people themselves using the European smallpox vaccination (ibid.: 45). Vaughan (1991: 24) also mentions how injections were internalized and 'indigenized', as well as other biomedical practices that were found useful. However, he rather argues that African healing systems were far from destroyed by the joint assault of colonialism and biomedicine; Prins (1989) also states that continuities with African ideas about health and healing are more remarkable than the fractures. What may, however, have weakened is a broader institutional organization holding these practices together.

As is the case with RCTs today, working together with governmental, ethical and legal organizations and standards to support their practices and evidence as legitimate, during this precolonial period, *muthi* practices were similarly in accordance with the political and judicial system in place. The judicial system of the Zulu kingdom consisted of two branches – the *ibandla* ceremony (king's council of elder statespersons) and the *umhlahlo* ceremony, which was presided over by *izangoma* (Flint 2008: 79). It was the second process, the *umhlahlo* ceremony, seemingly more opaque and arbitrary, that was eventually banned by British colonial administrators. The *umhlahlo* ceremony dealt with most cases of death and sickness, often involving important subjects and threatening political disorders, suspecting *umthakathis* (witches) or *idlozis* (ancestors). When sacrifices of the *idlozis* failed to bring about the expected recovery of the patient, people began to suspect poisoning or witchcraft (ibid.: 59).

A typical national *umhlahlo* ceremony could take up to four days, leaving time to summon *izangoma*, who often travelled long distances to attend. Issues of objectivity and fairness led to a preference for healers unfamiliar with the case at hand. Each doctor came with a group of armed followers to protect him or her from those involved. The *umhlahlo* began as the disputants and representatives of the community gathered in a circle. Each *isangoma* then entered the circle separately in an effort to divine the nature of the problem, with the most celebrated *isangoma* entering last. Ideally, this method was intended to keep the *izangoma* from knowing what the other doctors had divined; in practice, however, this was not always the case. The results of either the *ibandla* or *umhlahlo* would then be told to the king, who acted as final arbitrator and determined a settlement or punishment. 'Those "sniffed out" as umthakathis were considered dangerous and unredeemable and often faced imminent death, particularly in the early years of the kingdom' (ibid.: 80). Crampton also reports that during the seventeenth century, powerful Xhosa diviners could be summoned to 'smell out' the person responsible for 'bewitching' (2006: 212), showing how the most 'immeasurable' sense was found to be the most precise.

While these large trials provide some insights into ways *izangoma* may partake in healing the body social, Flint also points out how the fragile balance between healers and leaders did not always maintain a collaborative status. In particular, neighbouring Xhosa healers are mentioned to have 'successfully claimed a power base independent of their rulers' (2008: 84). In fact, various strategies were put into place to limit the power and influence of certain healers. The leaders themselves claimed to be powerful healers, often limiting who was allowed to practice by incorporating powerful healers into their elite. With the fall and disintegration of the Zulu kingdom, *izangoma* both gained autonomy and saw their political power and legitimacy diminish. The histories of the Zulu kingdom do not necessarily reflect those of Xhosa *izangoma*, yet they can tell of another social arrangement that did ensure some control over forms of healing. Acting as arbitrators of justice, *izangoma* were able to gain autonomy from rulers. This may help understand in part current practices of *izinyanga* and *izangoma* today, as they seem to operate under their own rules in their neighbourhoods. African healers were criminalized in 1862, and it is impressive to witness their still-important role in the townships of Cape Town today.

Wreford's (2008: 160) powerful testimony of her own passage into the role of an *isangoma* in 2001 shows how the *umhlahlo* is very much in place in the township of Khayelitsha. Even if not at legal national level, Wreford attests to the solid jury of *izangoma* that is mobilized to rule on the authenticity of an initiate through what she calls 'ancestral evidence' before passage as a healer is accepted. She not only shows how this passage was arduous for her as a white foreign woman, but also for her mentor, who is from Zimbabwe and thus not considered part of the local *isangoma* community. What is even more fascinating is the 'mystery' *isangoma* who is the real judge of her acquired abilities as a healer. As in Flint's historical account above, she is an *isangoma* who comes from elsewhere and who Wreford had never met, thus ensuring the 'objectivity' of the ruling.

What astonishes me in Wreford's account is the contemporaneity of these practices, while Flint's account speaks solely to precolonial times, yet most of what I heard in the Cape during my stay was a need to find a way to eliminate 'quacks'. None ever spoke of the solid courts of evidence telling who the 'true' healers are and who are not by reading the performances of these healers over the course of decades. The whole process of traditionalization of indigenous medicine seems to completely obliterate these practices, or perhaps they are meant to remain secret and only occur within a community of experts. In either case, when Wreford states that her role as a white *isangoma* may be to bridge Xhosa healing practices with those of biomedicine, I think she is definitely one who can do so, namely, by making visible the solid organization in place that decides who is skilled in healing. My own potential is weak compared to this; however, I am on a similar quest, perhaps only skimming the surface, yet through time and space I will hopefully also contribute to mending practices that have become so disparate, overshadowing each other rather than enlightening us regarding different healing routes and roots.

Beyond the legal route, more subtle forms of colonization can be understood as being done through Western medicine. In South Africa, missionaries tapped into the practice of healing the body as a way to act upon people's hearts and souls early on. Already in the 1890s, 'medicine became a potential gateway to changing African attitudes and knowledge of the body, illness, and misfortune' (Flint 2008: 95). Christianity and its missionaries were highly instrumental to the British colonial state, playing an intermediary role between African chiefs and white authority (ibid.: 110). Even to date, Christian songs and ancestors were called upon in healing initiations I observed in 2008. Not very different from the link between *muthi* and ruling in the Zulu kingdom, medicine and Western rule are ways of governing populations. This strategy has become even more central as new, sophisticated knowledge of universal biological bodies is today at the very core of human rights and humanitarian aid, knowledge gained through RCTs, which both legalize and moralize medicine. Foucault demonstrates this exquisitely with regard to the modern era in the West, naming it biopower (1976), as mentioned above.

Scholars have since brought some critiques and called for new adjustments of Foucault's theory of biopower. Rabinow (1996), for instance, found new forms of biopower to be more biosocial, implying a looping effect from social desires into biological undertakings. Fassin (2006a) found that biopower has moved from politics *on* life to politics *of* life, and Lock and Nguyen (2010) contend that Foucault's assessments were almost exclusively inspired by his experience of modernization in Europe, in France in particular, and need to be rethought if they are to be applicable elsewhere. Vaughan (1991) does precisely this to explain what might have occurred differently in Africa. He brings four subtleties to Foucault's theory of biopower, of which the fourth is particularly relevant for my topic. The first nuance he brings to the African case is that colonial states were not 'modern states', thus still relying upon 'repressive power'. The second is that the medical power/knowledge complex was much less central than in the European modern state. The third nuance, linked with the previous two, is that operating through individual subjectivities was much less opportune in Africa; the classification of groups or populations upon which one could act was thus a far more important construction than individualization. The fourth, and most relevant here, nuance in the African case is that '[b]y relying so heavily on older modes of production for its very success, colonial capitalism also helped create the discourse on the "traditional", non-individualized and "unknowing" collective being – the "African", a discourse to which the idea of difference was central' (ibid.: 11–12). This assessment corresponds to Leslie's (1980) critique of indigenous knowledge discussed previously, stating that such a concept emerges within modern forms of biopower, attesting, however, to the particular way in which these powers emerged in South Africa – in this case, in a politics of repression as well as of inclusion through exclusion, in which the people are sometimes asked to be productive workers and docile bodies, other times asked to revive their 'traditional knowledge', as partly done with Xhosa *izangoma*, and in still other instances asked to comply and standardize according to biomedical rationales.

In South African histories, it is Xhosa healers in the Cape Colony who posed the greatest challenge to British colonialism in 1850–53 (Flint 2008: 106). Prophets appealing to the ancestors, showing their ability to mobilize public opinion and act for anticolonial purposes led numerous Xhosa uprisings against colonial rule. The idea of introducing biomedicine to the African population as a means of governance nevertheless took hold strongly throughout much of the twentieth century, although always in high tension with older forms of governance (Vaughan 1991). Three main reasons were initially invoked to pursue biomedicine's entry into African society: one was to replace African healers with biomedical doctors as a means of political power; another was to combat 'superstition' through a rational approach to the body, health and wellness that discredits African healers; and the other was to tend to African health, particularly as it affected the white population (Flint 2008: 119). The criminalizing of *izangoma* and the decision to license only *izinyanga* were related to political concerns with the opacity of the work of the first, as well as their symbolic and effectual power (ibid.: 127). *Izinyanga* are far less threatening, and initial colonial laws forced them to work alone, splitting the African healing community (ibid.: 129). One of the unexpected effects of licensing *izinyanga* was to enable them to travel throughout the country as traders.

The visibility of *izinyanga* in the cities confirms that competition between biomedicine and *muthi* is in full tension. 'By the 1920s not only were *inyangas* perceived as a commercial threat by a number of biomedical practitioners, but so were the licensing and commercial development that had lent *inyangas* medical legitimacy in the eyes of both African and white clients' (ibid.: 141). Attempts to organize medicinal plant markets were at the heart of some of the meetings I attended in Cape Town in November 2007, showing how this 'competing' medical knowledge is a very current issue, which was certainly also revived with the new global market interest in 'wild plants' as well as in 'indigenous medicine'. Some *izinyanga* add chemicals and patent medicines to their businesses, and some shops in the townships advertise themselves as a 'Chemist Shop' (see figure 3.17).

Issues of regulation and standardization remain prevalent. In fact, the professionalization of *izinyanga* began as early as the 1920s. In order to face these legal challenges, class and status needed to be established; a 'formal' education was developed; and clear definitions of their practice versus that of 'the *isangoma* or "witch doctor" were produced' (ibid.: 150–54). In the earlier twentieth century, some African *izinyanga* professionalized their practices, seeking legal recognition for their organizations. Today more than 250 healing associations exist, 'who like their forebears are still negotiating with the South African government for legal recognition' (Flint 2001: 221).

The influence of Indian herbalists might finally be mentioned, as they have had quite an important influence on African herbal healing practices. Yet, as in the case of Rastafarian *bossiedoktors,* they are less often assigned the status of holders of 'indigenous knowledge'. Indian healers and *muthi* shop owners are instead assumed to be purveyors of goods by predominantly African policy makers. This is also the case for Rastafarians; they are even perceived as such by Xhosa healers. Both Indian and Rastafarian herbalists (and certainly Malaysian and Chinese herbalists) are seen as foreign and do not fit into the category of 'indigenous knowledge', which is mainly reserved for native tribal Africans. With regard to Rastafarian *bossiedoktors* and Xhosa *izangoma,* Indian influences are, however, relevant in both cases. Brahman traditions were brought to the Caribbean in the twentieth century via the migration of many thousands of Hindus to Jamaica, the birthing place of Rastafarian traditions (Black 1992: 89). In the South African case, more than one million Indians emigrated to the Natal region at the end of the nineteenth century (Flint 2006), also possibly bringing some aspects of Brahman traditions; numerous Hindu and Indian priests of all castes became herbalists, collaborating with both Rastafarian *bossiedoktors* and Xhosa healers. This might explain some of their commonalities or the common struggle seemingly shared by Rastafarian and Xhosa healers. Rastafarian *bossiedoktors* fell through the cracks and are currently claiming their status as African healers, while Xhosa *izinyanga* and Xhosa *izangoma* began some forms of professionalization in the postapartheid era.

Moving In

This chapter began by moving a little bit deeper into issues of indigeneity, in particular within its politics and its histories broadly and, more specifically, in those histories playing themselves out within the contexts of the preclinical trial. I have paid attention to the politics of indigeneity in anthropology and as they emerged within the NC-CAM and the IKS branch of the MRC. I then evoked aspects of precolonial histories that appear to emerge in current practices or help understand how current practices might make precolonial histories appear differently today. I also addressed the emergence of *muthi* as a broad category to include multiple forms of indigenous medicine which are unrecognizable as biomedicine in postcolonial times, as well as how this connects to *inyanga* and *isangoma* practices within the South African nation. The following two chapters move in more depth into 'indigenous healing practices' as I was able to engage with them during my time in the Cape: respectively, those of Rastafarian *bossiedoktors* and Xhosa *izangoma*.

❧ Chapter 5
Healing the Nation

One aim, one destiny, what we live for is *inity.*

– Everlivings, lyrics from a song recorded in 2008

The South African peninsula assembles one of the largest groups of Rastafarians in the world, especially the regions of the Northern Cape, Western Cape, Gauteng, Free State, Limpopo and North West Provinces in South Africa and Lesotho, Swaziland Mozambique and Botswana (Mantula 2006: 1). The Rastafarian live according to ancient African roots meant to heal the past through immediacy. With them I learned from their practices with plants, including *A. afra,* and from the centrality of music in healing. I learned how they collaborate with Xhosa *izangoma* as well as occupy the markets selling herbs. The roots of the movement and the ways it is currently lived in the Cape help with understanding issues of dignity running through the preclinical trial and at its edges. First, I describe how my family and I became 'part of the struggle', namely, the Rastafarian struggle of mending a wounded world. Second, I show how the roots of the movement in Jamaica are lived in the Cape and how they fit within South Africa and Africa more generally through KhoiSan roots. A third section discusses how sounds and a particular 'way of being' or *livity* ground Rastafarians' ways of healing with plants in a link with the earth and a desire to retrieve and establish the dignity of a people, thus aiming to heal the nation. In a fourth section I discuss the place of herbs and of the sacred Herb in healing. Finally, I discuss plant sentience as shared across healing practices in the peripheries.

'Part of the Struggle'

We were welcomed in the home of a couple of Rastafarian *izinyanga*, or *bossiedoktors* (bush doctors) as they also refer to themselves in Afrikaans, on a small parcel of fenced land by the side of the highway in Delft township (figure 5.1); they and their six children shared a two-room cement house. Two adjacent tin houses sat nearby: one was used to conserve the plants collected, the other served as a music studio/practice room. My husband, a professional musician, who was accompanying me on that day, disappeared in the studio following our visit to the plant stock room. Back in the central house, I noticed how the relations that I had established with the *bossiedoktors* during the previous months seemed to have drastically changed. While beforehand they had collaborated in a collegial fashion with my interests in their use of medicinal plants, in discovering that my husband dedicates his life to music and knew reggae rhythms, they suddenly adopted us as 'part of the struggle', meaning the Rastafarian struggle towards a better world through *livity* or a particular way of being in *inity* (unity). Our investment in making music made us part of their way of living and of knowing life, partaking in their solution to current issues of injustice, inequality, poverty, violence and so forth. With this connection through music, the host *bossiedoktor* and his wife explained that I could learn how they know the benefits of plants because, as all knowledge, it needs to be felt, and this is done through music, listening and moving according to certain rhythms.

Figure 5.1. Delft township, Cape Town, February 2008

It was under these circumstances that the process of musical creativity appeared as a way to acquire the necessary abilities to be able to know how to use plants to heal, very much as the Xhosa *isangoma* indicated in inviting me to a drumming session. Beyond the plants, these particular relations with the world refer to the ability to be, to live a certain way. Creating music, listening and performing are part of the desired *livity*. Making music is a source of embodied knowledge and the power to heal wounds or (dis)ease for those partaking in making a better world. Making music is about living in the everyday a certain way. Words are central, and they are to be lived as well, words and acts needing to match to maintain a good life.

The family of herbalists/musicians named their musical formation the Everlivings (figure 5.2). The father converted to Rastafarianism a decade earlier in order to escape the violence of street gangs in the townships. The Rastafarian way of being, based on a philosophy of nonviolence, is what originally attracted him to the movement. The musical project, he explains, is a family project, a way to keep the family together, to spend time with his children, which he did not have a chance to do with his own father. It was also explained as a way to keep the children 'out of trouble', as his wife explained. Discipline, art and hope are hence linked to this musical formation, which features the two young brothers on vocals, the older brother on the synthesizer, the other on percussion, the sister on the bass, the father on drums and a friend of the family, of Malaysian origin, on the guitar. Their maternal language is Afrikaans, yet the lyrics composed are in English, one of the eleven official languages in South Africa as well as the language of the Jamaican Rastafarian tradition. The musical style is reggae, a style developed in Jamaica that plays three notes out of four; the silent note is to be felt rather then played, making another strong statement linked with *livity*. The words are taken from the psalms of the Bible in line with the Rastafarian tradition and are intimately linked to life lived in the townships, often from the point of view of the youth (aged between ten and twenty-one at the time), as in the following lyrics:

> We are the children of this world. (3X)
> We are the leaders of tomorrow.
> We let the fire, fire burn. (3X)
> Let us be what we want to be ...
> Please stop this war against Jah Creation.
> People unite, stop rushing and fight.
> Time to see Jah light, shine so bright. (2X)

Another song goes as follows:

> Some say I'm just a child.
> What do I know about life?

Figure 5.2. The Everlivings, photograph by Melissa Robertson, February 2008

Jah say, Jah know.
I'm a Jah son.
All I want to say.
Wonder for my nation.

Their musical performances move through various locations of the townships and festivals across the country, sometimes more political and accompanied by oratory art and transmitting more specific messages, other times more musical, featuring new musicians on stage. In contrast with *isangoma* performances, which are levelled to the ground and have everyone participate, Rastafarian performances use stages and face the audience. The stages are, however, filled with people and families from the audience; they are neither above nor distant or separated from the crowd. I was surprised at one of the festivals to find the stage lower than the audience, making it clear that it is music for, with and about the people, coming straight from the mouth of children, in this case. Yet another song began in the following way:

Once a man and twice a child
What we do is for a while
Today for me, tomorrow for you
The harder they come, the harder they fall
Peace and love in your neighbourhood (3X)
We got to rock to the rhythm of the rastaman
We got to rock to the rhythm of the fireman
Can you feel this fire, feel it in your soul
This is my desire to let the people know
Let the truth be told, doesn't come a day gone so
Young and old, Selassie never leave me alone
It doesn't matter where you're coming from
You gotta rock to the rhythm of this fireman if you come from Jamaica (2X)
Open up your eyes and look at all the things are going down
Doesn't matter who you are, you have a right to know what life is all about
Because everybody is dying, all the time is flying
We've got to keep on trying, cause we struggling and dying here
Peace and love in your neighbourhood (3X)

Since my husband became 'part of the band', we undertook a recording session, which was set in the 'open air' in a tin garage filled with transmitters in the middle of the township of Hout Bay with many listeners from the neighbourhood sitting closely to listen.

The songs, written in the townships and about life in the townships in postapartheid South Africa, provide a sense of the importance of the Rastafarian struggle for liberation from oppression and racial discrimination. Through music, the everyday struggles and ways to live through them is expressed. In the opening words of one of the songs, the epigraph for this chapter, the words express a vision of a people, *inity* replacing 'unity' to refer to the necessary union with Jah. The introduction to another song again alludes to this and expresses that this union, or the transcendent love of Jah, is liberating: 'Jah love, set us free from misery, slavery, captivity, poverty.' This liberty helps some detach from the oppression, corruption, sins and vanity linked to Babylon and against which the movement built itself in Jamaica.

What is promoted in the Rafastarian songs, written and performed by this family, is equality for all, namely, with regard to the possibility of 'knowing' (in the sense I define in chapter 1), as these lyrics express: 'It doesn't matter who you are. You have the right to know what life is all about.' Such lyrics allude to forms of monopoly of knowledge by oppressors or colonizers, which can be linked to the deception felt in the visit to the Delft laboratories facility, the feeling that what was happening in the laboratory had been hidden from them (as discussed in chapter 3). In a sense, they point to the inability for all to benefit from research done within such laboratories, let alone have the impossibility of utilizing the technologies to transform plants into pills, powders and capsules.

Taking care of each other is another strong message, as found in lyrics such as: 'Don't be late for your brother/sister. Someday you will need each other.' Another strong message is to take care of the world, as found in lyrics like the following: 'Jah created us so that we can understand each other/love each other' and again, 'Please stop this war against Jah creation.' *Livity* is all of these things, as they should be enacted in the everyday.

These sung words are shaped by everyday life as well as continuously shape everyday acts, including the ways wild plants should be collected. Ways of doing so often involve meditative walks and collecting plants by hand respectfully and without greed. It is impressive to see that in quite rough living conditions, these 'ideals' are fully lived up to in how the practitioners enter into relations with both people and plants. Eating raw vegetarian foods in conditions in which meat is easier to obtain is one of these challenges that is up-held, even at the cost of not eating at all should these not be found. The challenges with regard to maintaining good relations with people are also responded to in priestly fashion. The son-in-law of the Rastafarian *bossiedoktor* described here was, for instance, shot twice during his work as a taxi driver in the townships; a violent aggression explained as stemming from competition and jealousy between two taxi companies. Luckily, the twenty-year-old received the bullets superficially in the shoulder and survived, offering complete pardon to his aggressors and continuing his job without fear. Similarly, it is the laws of Jah that prevail in their work with herbs.

Connection to herbs through the Creator are found more legitimate than the laws that stem from Western capitalism and imperialism to which they do not adhere; namely, the laws permitting to close off lands as private property or as areas of conservation for parks, gardens or tourist activities. It is in this way that life is known and life in its greed and excess is rejected, and hence will not be enacted. This is known by most people in Cape Town, who generally show genuine appreciation for the Rastafarians' 'priestly' practices and perhaps as well for their healing skills. Knowing life in the 'here and now' and maximizing the moment invites one to dwell in immediacy or in the intensity of a continuously emerging world that we need to meliorate through every act.

Roots of the Movement

> The experience of the openness of life, of the common origin of all creation, is a primordial understanding, which according to the brethren constitutes the source of their spiritual energies.
>
> – Carole Yawney, 'Lions in Babylon'

The Rastafarian practices in the Cape show local colours, but they are also situated in the global Rastafarian movement that emerged in Jamaica in the 1930s. The inspiration of the movement is African, originating in the Ancient Order of African Ethiopian life of Ras Tafari, yet emerged in the Caribbean from a situation of oppression and poverty in direct confrontation with colonial powers. Marcus Mosiah Garvey, the founder of the 'return to Africa' movement, is considered to be the prophet of black liberation of the twentieth century by the Rastafarian movement. The Rastafarian community in Philippi township, where I spent time with Rastafarian families in the Cape, is named after him (figure 5.3).

Born in Jamaica in 1887 and having lived with racial discrimination all his life, Marcus Garvey later articulated a theological basis to counter racial discrimination. His contribution was in highlighting the problems of discrimination and proposing an acceptance of racial difference by calling for dignity and justice for all (Eskrine 2005: 31). Leonard Howell is considered the founder of the Rastafarian movement, which he institutionalized in Kingston, Jamaica, in the 1930s. At this time the world witnessed the coronation of the Ethiopian emperor Haile Sélassie I, who exposed an ancient African ritual to the world scene for the first time, a ritual that realized the Rastafarian prophecy – that Haile Sélassie is the reincarnation of Jesus Christ, from the filiation of Solomon and Adam, thus bringing Africans to the source of humanity as found in the *Kebra Nagast*,[1] the lost Bible of Rastafarian wisdom and faith from Ethiopia. Prince Ras

Figure 5.3. Marcus Garvey Rastafarian community welcoming banner, Philippi township, Cape Town, January 2008

Tafari became the sign of the manifestation of Jah (God) and Ethiopia the Zion (Jerusalem) of the black man. The movement 'emerged as one of the most articulate alternative philosophical paradigms to modern capitalistic imperialism' (Niaah 2003: 825).

Carole Diane Yawney, whose quote opens this section, was an assistant professor in medical anthropology at York University who lived with the 'brethren' and 'sistren' in Jamaica from 1970 until her passing on 23 July 2005 – namely, the 113th anniversary of the birth of Emperor Haile Sélassie, on the same day Ras Mortimo Planno, her mentor, had his right leg amputated due to health complications. She played the role of secretary of international liaisons for Planno. Planno was a Rastafarian leader who welcomed Emperor Halle Sélassie when he disembarked from his plane in 1966 and was central to the movement. These coincidences in the dates of 'happenings' are important within the movement, read as 'signs' of movements forwards in the wanted direction. My own birthday coinciding with the day of the birth of the emperor on 23 July, together with my husband's birthday coinciding with the day of the coronation of the emperor on 2 November, were, for instance, taken as further signs of our being 'part of the struggle'. Yawney's connection to the movement as Planno's apprentice made it so that she played an active role within the movement itself. During her thirty-five years learning with the Rastafarians as well as fuelling their liberation movement, which was expressly anticolonial in outlook, Yawney took the position, unpopular among many Rastafarians, that patriarchy within the movement was a remnant of a colonial mentality. Her role in making a place for women within this liberation remains overshadowed. Mostly men become 'priests' and lead the movement; however, 'sistren' were also part of the band discussed above, and numerous Rastafarian women were active *izinyanga* in the Cape. In her journey with the Rastafarians, Yawney ultimately travelled to Ethiopia and South Africa, where the movement was taking root.

The heart of the movement, to which all Rastafarians adhere, concerns sharing a common sense of 'good' and a common sense of 'identity/solidarity' (Edmonds 2003: 67). The 'bad' is incarnated in the term 'Babylon' and consists of the cultural and political power of the whites through colonialism, imperialism and racism. In the Christian world, as in the Rastafarian world, the myths related to Babylon retake the topos of the apocalypse, an excess of sin and vanity leading to the downfall of a civilization. The term 'Babylon' is used today in the Rastafarian movement to warn against the abuses of the capitalist world of excessive consumption, a statement that touches upon ways to manipulate, sell and consume plants. The proposition of the movement is to surpass all judgement of race and class and to rebuild a world of solidary common sense. This means reestablishing the dignity of the discriminated people by going back to their African roots for liberation of oppression and colonial injustice. The Rastafarian alternative theology survives today and is in expansion. It constitutes an attractive movement on a global scale, a particular way of being, a modus vivendi of which the essential elements remain informal networks of communication, a shared ideology and common ritual practices (ibid.), all of which have found a central place in South Africa.

The Rastafarian movement is increasing in scope in South Africa, permitting some to live a life away from violence in the gangs of the townships and others to surpass the lived experience of racial discrimination. It is a movement that currently fits with the aspirations of African dignity in postapartheid South Africa. South Africa, with its recent history of 'decolonization', exhibits renewed respect for the Rastafarians in the everyday for their role as keepers of social justice, a role they played during the apartheid. Rastafarians played the role of guardians of peace during the South African apartheid, their herb stalls serving as a refuge for the passage from black to white zones (and the reverse). In the stories I heard and reheard from both Rastafarians and non-Rastafarians during my stay in the Cape, for example, people hid behind the stalls at the arrival of Afrikaans police. This role as guardians of peace is preserved in postapartheid South Africa. For example, in a settlement being rebuilt in the townships of Cape Town, 'leaders allocated a Rastafari-headed household to each section of the new settlement. Rastafari were expected to act as calming influences and to mediate potential (racial) conflicts' (Ross 2010: 172).

Most Rastafarians speak Afrikaans, the language spoken by the Dutch colonizers whose descendants are known as the Boers or Afrikaners; it is a language blend that was first written in Arabic script and that mixes languages spoken by slaves of all nationalities that were brought to the Cape by Dutch colonists – Malagasys, Mozambicans, Indians, Malays, Angolans and so forth. The Afrikaans language evolved from a kind of kitchen Dutch blended with elements of all of these other disparate languages, as well as that of Khoi 'apprentices' (Crampton 2006: 295). A power struggle between the English and Dutch colonizers culminated in a majority for the Afrikaners' National Party in the 1940s, and it was this party that put into place the apartheid laws in 1948, laws that institutionalized racial discrimination. Under the laws of racial discrimination, South Africans were classified into three categories: the whites, the blacks (Africans) and the 'coloureds' (of mixed descent). The category 'coloureds' included the major Indian and Asian subgroups. The Rastafarians of the Cape are of mixed descent, and most were classified as 'coloureds' in the apartheid hierarchy, a hierarchy strongly resisted by the Rastafarian movement. The laws of apartheid were officially banned in 1991. This recent history of the South African apartheid makes the territory prone to the expansion of the Rastafarian movement, as it fits particularly well with this moment of the 'African Renaissance'. Rastafarian bush doctors, as Kroll names them (2006: 241), refer to KhoiSan roots as the source of their 'authentic African' knowledge, knowledge that would be transmitted by coloured elders. The KhoiSan were the 'first inhabitants' of the South African peninsula, also known as the Kalahari Bushmen, San Bushmen or simply the San, replacing Jamaican Rastafarian's Ethiopia with the Kalahari. The Rastafarian movement is thus a movement of return to these roots, brought into being in the everyday through words, visions, sounds and acts, all done through a particular 'way of being' or *livity*.

Livity and Word-Sound-Power

Livity is the way of being in the world as named in Rastafarian 'I talk'. Living, feeling, is knowing; 'knowing is always characterized by a high degree of certainty, by a close relation to the practical, and by an innate presence within man' (Yawney 1978: 215). It is about living in the present transcended by Jah (God). 'To know' opposes itself directly to 'to think' (Gratton 1986: 191). Knowing is further distinguished from 'hearing', 'learning' and 'reading'. 'These are all modes of inducing belief, not knowledge, and were used by the colonial masters to enslave the people's minds' (Yawney 1978: 215). 'To know' is knowing by feeling, not believing from afar. While sounds are ways of knowing, they need to be felt. Learning is living. Reading is knowing only when embodied through visions. Vision is also received and embodied. Seeing is knowing rather than believing, since it is felt, as in seeing 'Jah light'; it is living in accordance with Jah as well as in accordance with a well-defined philosophy of justice for all. These evoked 'ways of knowing' echo with those of the Xhosa *izangoma*, as well as with Ingold's (2013) recent distinction between knowing from the inside as opposed to 'information', as discussed in chapter 1. It also corresponds to certain insights from phenomenology in opposition to operational knowledge, as proposed by Merleau-Ponty (1945); however, the Rastafarian case shows more directly the political implications of distant ways of knowing, as they disempower all but the experts, who obtain them through formal education. This recalls the hierarchy between primary and secondary qualities as enunciated by Latour in chapter 1's epigraph – delving into the very secondary qualities of the visible, the lived and the felt as ways to surpass the claim of knowing what the world is really made of (one nature) in order to partake in making a common world.

This is *livity*, or simply 'to live', and it is certainly opposed to the common ideas of perception embedded in the RCT. *Livity* incarnates the theories of perception developed by Ingold and Merleau-Ponty. As is the case with these thinkers in the phenomenological tradition as well with the *izangoma*, the preobjective is highly valued. This might further explain how Rastafarian *bossiedoktors* and *izangoma* agree with each other's ways of making plants work. Contrary to the *izangoma*, however, Rastafarians work more directly with the plants in their materialities. They sell them in stalls as commodities, side by side with socks, cloths or other items (see figure 3.15). Plants are collected in the 'wild', dried and tied into bundles and sold in markets and festivals. Some Rastafarian *sakmanne*, who dedicate seven years of their life to living without material goods, wearing a single cloth, walking barefoot and eating raw foods given by the community, sell wild plants and roots during festivals (see figures 3.8 and 3.20). Most Rastafarians are vegetarian and entertain particular relations with food, which is given directly by Jah, the Creator. Rastafarians do not consider themselves to be masters of 'nature', but rather sustain a particular relation with plants as given by Jah. This is all part of *livity*, and to live in this way is to know.

Words and sounds are central to Rastafarians' way of being, to be felt, embodied. Music is central to *livity,* as stated by Bob Marley's son: 'My father's music lives as he lives in the music.... The music comes from Jah.... [I]f a song makes sense, it will always be there for people to understand' (Marley 1997: 7). The Rastafarian movement strategically plays upon the power of words and sounds for its expansion, possibly selecting consciously those very senses underprivileged by colonial powers. Oratory art in the streets of Kingston in Jamaica shared the words of Jah with all who desired liberation from colonial oppression. This 'art of saying' transformed itself into 'musical art' at the heart of the movement, as well as enabling it to travel. The message is taken into musical expression through music for the soul; misery is put into melody through various forms such as burru drumming, kumina, nyabinghi chanting, reggae, dancehall, calypso, soca and dub (Niaah 2003: 832). Reggae developed in the 1960s and became the most known of these musical expressions internationally. Reggae can be recognized for the rhythmic accent placed on the offbeat. Its tempo is slower than other popular forms such as Jamaican ska and rocksteady. An initiative of Island Records Studio to accommodate Bob Marley's music to the American ear gave the movement power to expand on a world scale; radio 'airwaves' is the way the movement is said to have arrived in South Africa, where Jah's message seemed to fit and proliferate. Similar ways of entering relations with nonhumans fit into this ontology, including ways of engaging with plants that are of prime importance in the larger struggle to heal the nation, all achieved through the power of words and sounds.

Some speak of the '[r]ise of Rastafari Livity in the "post apartheid" South Africa, preferably called Azania by some tenets of the movement' (Mthembu 2007: 2). Rastafarian 'I talk', or the 'I and I' philosophy, is with direct reference to the transcendence of Jah (God) in each and every one of us. 'I and I' refers to

> both first person singular and plural, subject and object. The context alone makes it clear to whom one is referring. The concept of 'I and I' both reaffirms the identity of the individual as an individual, while at the same time stressing his union with other beings and the forces of life.... The sound 'I', which is also associated with the idea of eye or vision, is combined in a number of ways to create word variations. It is often substituted for the sound 'You' or for key syllables in a word. (Yawney 1978: 219)

All actions in the everyday are transcended by Jah, including ways of 'being with' what Jah created, including plants in healing, of which one is sacred, as I will discuss in the following section. Others are valuable for their 'wildness', as they are closest to Jah, who created them.

The Rastafarian *bossiedoktors* I met were either musicians or highly attuned to musical expression; the two families I knew most took me to numerous festivals and events in which we were immersed sometimes for days. I accompanied Rastafarians to more than ten festivals, thus meeting large groups of Rastafarians in the process. Wild plants were always part of the picture, with *sakmanne* sitting amongst them for sale (see figure 3.20). While sounds are not explicitly central to organized phased ritual healing as in *izangoma* practices, playing, listening, dancing, moving with Rasta music and receiving Jah in visions through forms of trance and meditation are ways of being. *Livity* implies certain ways of moving, vibrating in accordance with reggae rhythms linked to the social and environmental justice philosophy. *Livity* also guides ways of engaging with and collecting plants, especially in regard to their efficacy and possibilities to heal. Creating, listening and performing music are central to this way of being-in-the-world, in this case a way of being as 'offbeat' as reggae itself, or rather a way of being 'in the desired beat', which is explicitly to resist imperialist capitalism. Creating, performing, listening and moving to these rhythms is a source of knowledge, power and a way to heal social wounds. Musical creativity in the everyday is part of the struggle to ensure the survival of this specific way of being-in-the-world. That I was 'part of the struggle' when I helped transport musical instruments to the stage expresses how *livity* is in everyday engagements with the world. And part of reaching the desired *livity* is the Herb and herbs.

Herbs, Herb and Healing

Knowing *A. afra* emerges through *livity*. The plant is not objectified. It is a creation of Jah and its multiple uses are those learned in life, from people buying the plant, other *izinyanga* and elders. The family living in Delft township kept an *A. afra* bush growing in their yard (see figure 3.11); they prepared it in teas and rubbed it on the stomachs of feverish babies with oil. They showed me their shack to keep the dried plants they had collected and tied into neat packets for selling in the markets, in the process telling me about all the other plants and roots as well, even *dassiepis* (petrified Cape hyrax urine), highly prized for stimulating fertility. It could be assumed that selling the plants turns them into a commodity and participates in Babylon's greed; however, the intentionality and the way it is done makes it distinct. Some selective forms of commodification seem acceptable; Rastafarians, for instance, buy plants transformed into tinctures and powders, yet solely those from Jamaican brands or with Rastafarian colours. The Rastafarian movement travels through global networks, or 'airwaves', accumulating paraphernalia, which tends to commercialize it as well, attracting numerous tourists; however, this is found to support the struggle. For those who live the movement thoroughly, of importance is that the particular ways of knowing life are those that can be enacted in the everyday, as well as those that can be envisioned for a better future. Knowing life at the molecular level is hence not a challenge to *livity* in itself; the problem is in the way it is done as well as how it is known and whether this knowledge is accessible to all (or not). Life and life in things is known, since it is felt, and letting this 'right' wither away is what becomes 'part of the problem', namely, injustice. To keep within the solution, I have shown how music, sounds and words ensure that every act is felt, and the Herb participates in keeping with this particular way of being.

The sacred Herb (*Cannabis sativa*, marijuana, *dagga* in Afrikaans) became an explicit part of the Rastafarian movement in 1940 with Leonard Howell in the Pinnacle commune (Niaah 2003) in Jamaica. The Herb is another token of resistance to movements of domination of the imperialist capitalist world judged corrupted. Illegal in South Africa since 1928 under the Medical, Dental and Pharmacy Act and under the Weeds Act since 1937, it became generally criminalized in 1992. Rastafarians are often arrested, and they pursue usage under religious motifs as well as a disagreement with man controlling the use of certain plants, especially a sacred one (Laplante 2009a: 115). While the Rastafarian movement cannot be reduced to the consumption of *dagga*, the Herb plays a key role in healing with plants. It plays a role on numerous levels, both material and immaterial. It is found growing on most parcels of land of Rastafarian families in the townships, side by side with A. *afra*. Dried leaves and flowers to smoke are conserved inside the hats of most Rastafarian men, chillum in hand. *Dagga* is often served on a silver platter and glorified by sacred words and ritual gestures during its usage, sale and acquisition, notwithstanding the materiality of the smell of its smoke embalming the air at proximity.

The KhoiSan people to which the South African Rastafarians look as a source for ancient African roots and knowledge on plants would, according to Low, place particular importance on smell: 'Smell is more commonly the way plant properties are conceptualized. It is for instance the smell of a plant that removes sickness from a body, whereas in treatments that involve human healing or animal medicine, wind or smell tends to be used to describe the active principle' (Low 2008: 68). The only two plants grown in the backyard of the Rastafarian *bossiedoktor* and musician family were *A. afra* and *dagga,* both aromatic plants and known perfectly well from their smells, although also from their textures, visuals and tastes. While using *dagga* is definitely linked with the transnational Rastafarian movement, it also has its roots in Africa; the term *dagga* is found in nonindigenous, social, medical and legal texts throughout the greater part of the twentieth century (Kepe 2003: 3). The use of *A. afra* amongst coloured elders, to which the South African Rastafarians also turn for knowledge of plants, was also found to be prevalent, as reported in two small surveys done with older coloureds in the Cape.[2] *Buchu* (*Agathosma betulina*) is the second most common herb, a plant to which I really took in my own everyday, as its tea is very aromatic, seems to thicken the water and provides a sensation of warmth.

The family in Delft preferred *A. afra* and *dagga* for everyday uses, the father of the family often referring to 'the Herb' as 'soul food', literally replacing meals, explaining it serves to refill the soul with hope and courage to pursue, in all integrity, with the hardships of the everyday struggle. The awareness of the environment that its smoke can provide if intended in this way is potentially relevant to the ways of engaging with plants that are deemed to provide efficacy. Further, the environment and the plant is given agency by the KhoiSan people, as with the Rastafarians: 'a suitable plant will go to the site of sickness because it smells it, or alternatively the smell of the plant takes out the sickness' (Low 2008: 77). Low's finding can be extrapolated to the therapeutic benefits of the odour of the sacred plant during social events, festivals, performances and everyday life;

however, as with *A. afra,* I rather found it to be a mixture of senses, synaesthesias, that were part of the benefits of these plants, a mixture of senses modified in relation to contexts and engagements with the plants.

Generally speaking, imbalance in life, or disease, is considered to take root in the separation of Jah from thought, in the separation of words and acts. The use of *dagga* facilitates access to certain forms of knowledge linked with the altered states of consciousness it causes, which are highly valourized as well as associated with wisdom. Reaching these states of consciousness allows the emergence of incredible intuitions with regard to the use of plants (Kroll 2006: 241). The use of *dagga* also has the potential to enhance one's ability to hear certain sounds, as also shown in classical clinical trials (Laplante 2012). This potential is enhanced by a combination of other acts in the everyday, in line with the visionary movement; an apprenticeship enables one to orient the desired effects according to these abilities and intentions (Yawney 1978: 16). As such, it is not useful to solely reduce *Cannabis sativa* to the level of a chemical molecule having a specific effect in itself. What doing so nevertheless demonstrates is that the active compound tetrahydrocannabinol (THC) found in the plant shows neurophysiological effects on auditory perception: 'it can help experimented musicians to play more intensively during improvisation' (Fachner 2006: 82), favours hyper concentration in the acoustic space and encourages a more effective attention on auditory information (Becker 1963; Curry 1968). According to these studies, cannabis acts as an agent of intensification, as an exciter, equalizer or psychoacoustic attenuator making sounds more transparent and the sources of sounds more distinct, and provokes 'a greater special separation between the sources of sounds and the perception of subtle changes in sounds' (Tart 1971). What these studies continue to omit, however, is how the perceiver can attenuate, enhance, tailor or ignore these effects as well. The mediums in which the Herb is taken, where it comes from and in which circumstances will also take part in the experience.

'The French poet Charles Baudelaire's (1966) and Tart's (1971) descriptions of synesthetic effects (a mixing of senses), weakened censorship of visual depth perception (Emrich et al. 1991) and a change toward a field-dependent style of thinking (Dinnerstein 1966) suggest an intensification of the individual cerebral hearing regime' (Fachner 2006: 82). Whether it is the plant itself that provides these abilities, as assumed in the clinical process, or its consumers that orient the plant to their desired vocation, Rastafarian intentions, *dagga* and music entangle in *livity*. The prophet in the Rastafarian movement is the one who can best interpret the general ideological understandings in a way that can inspire others to creativity and innovation (Yawney 1978: 102), and this is also played out through music. The art of saying through sounds relies in part on the potential to orient creativity, the latter being enhanced through a channelling of the states of consciousness that can be facilitated by *dagga* (or not), moving beyond auditory functions and linking itself to vibrations and visions anchored in particular ways of being-in-the-world.

Practices surrounding the use of *dagga* thus play a role within healing. Cannabis amongst the Rastafarians of the Cape plays a part in the definition of roles, genres and statuses, as it is mainly priests who consume, more rarely women and children (I only became aware of one female priestess during our stay; she was of Jamaican origin, and that may explain her respected status, as with all things coming from the origins of the movement). *Dagga* also plays a role in defining identity through its explicit links to the movement. It symbolizes an identity of marginality due to the illegality of the plant, yet also one of pride of resistance, of integrity and social cohesion, forming links of trust and belonging from this positioning. According to Yawney's observations in Jamaica (1978: 75), the Herb is consumed as a source of social and spiritual healing, a source of illumination. In Cape Town, I noticed how it served meditative purposes, often in trips to the mountains to collect plants, either solo or in a group, aiming to renew links with 'roots'; *dagga* facilitates 'the revelation of oneself' in the world as well as relations with the world. These meditative states, as well as hardships, translate into ways of saying through musical expression, which is the medium towards a better life. Meditations are played out in life, as they enhance knowing life; musical expression is a way to share these meditations. Meditative use of *dagga* demonstrates the Rastafarian philosophy of social justice, of everyday struggles and *livity*. These explicit features of the movement are key to understanding healing with plants in the everyday, which also entangles with yet another kind of life to which both *izinyanga* and *izangoma* attend and that escapes the preclinical trial, namely, plant sentience.

Plant 'Wild' Life

'Life' activated in plants by healers relies on both the healers' abilities as well as on the plants' potentialities and histories in relations with humans. 'Life' in molecules in science is confined to the human body, trusting scientific procedures to finds ways in 'nature' to enhance it. The current frenzy around 'wild' plants in Europe and the United States might point in a similar direction. The very notion of 'wild' refers to plants out of the control of the mastery of humans. It evokes a naturalist imagery of the world as separate from humans. This is similar to the distinction between 'wild' animals and domesticated animals (Ingold 2000: 62). The 'wild' qualifier is not used by healers, as far as I know, and it is not because the plant is 'wild' in the sense of the former that it is more useful; rather, it is useful because it has the vital force or strength that is allocated to the bush, while the cultivated plant is not useful because the vital force is used up: 'Knowledge dissipates … and power evaporates unless reinvigorated from the bush', as van Beek and Banga mention with regard to Dogon cosmology (1992: 69; see also Ingold 2000: 84). For the Rastafarian it may be that it is Jah's creation and knowledge (as attuned attention in the world) that is used up; for the Xhosa healer it is the connections with the ancestors which are broken. In both cases, it is the plant's vitality that dissipates in cultivation.

In this respect, the Rastafarian movement may correspond in part to Indian practices, which are highly influent in the Eastern Cape *izinyanga* trade where they often travel. Ayurvedic medicines in Hindu parts of India are said to come from Brahman (God) the creator while for Rastafarians they come from Jah (God) the Creator who gave his knowledge to those closer to the roots of humanity, in South Africa becoming those of the KhoiSan bushmen of the Kalahari desert. Other Rastafarian practices can also recall those of the 'world renouncer' or Brahman found in Hinduism. Rastafarian *sakmanne* (bag-men) and *kaalvoetmanne* (bare-feet men) are two smaller segments within the *bossiedokters*, considered 'the most spiritualist members of the Rastafarian community' (Olivier 2011: 45).

'Although RasTafari lifestyle already involves the abstinence from things of 'Babylon', Bossiedokters – and especially the Sakmanne – are considered to be the strictest followers' (ibid.: 46). They classically walk barefoot to feel the soil with only a burlap sack for dress and live off wild plants. I've been told by many Rastafarians that they have been *sakmanne* at some stage or their life, a stage which consists in a seven year meditative apprenticeship which some do for longer. Rastafarian *sakmanne* are particularly present in the Cape, establishing intimate relations with wild herbs, barks and roots upon which they rely for their livelihoods and for those of others. *Sakmanne* live from wild plants, roots and minerals (see figure 3.19). The Rastafarian diet is called Ital or 'I-tal' based on the English word 'vital', condemning meat and alcohol, yet also 'artificial' food, or food which contains conservation agents or simply wrongful manipulations of the food in its preparation and thus making it loose its benefits for humans.

This appeal to vegetal life further coincides with KhoiSan roots which the *bossiedoktors* aim to perpetuate as well,

> whereas pills are objects, medicinal plants are much more. They have spirit or essence and are linked or entangled through the movement of *máq* – air, breath, wind, energy and vitality – between them, over time and through space (Low, 2009). Nani, an old and experienced healer tried to explain: When you walk, you leave a *spoor* (a trail), animals leave a *spoor*, rain leaves a *spoor*… you can see from the *spoor* what kind of rain was falling, hard rain, soft rain. Even plants leave a *spoor* – the grass drags on the ground, it falls in places, the seed falls, it rolls, the leaves, fall and blow around. You can see it was there. You leave a *spoor* and the leaves or the grass or seeds, they blow over it, then you can see you shared that story, that place even long after(wards). The *máq* moves between us, we share it. Plants are alive, they have a power to live, they change the soil around them, they can affect animals, they protect themselves from humans from animals, they move their seeds, they can *trek* (migrate) to other places and grow there. (Gibson 2011: 56–57)

While the histories of clinical trials are not devoid of experiments giving plants sentience, the latter remain peripheral and perceived as a disturbance. For instance, Backster's (1968, 1973) 'discovery' that plants appear to be sentient caused strong and varied reaction round the globe, despite the fact that Backster never claimed a discovery, only an uncovering of what has been known and forgotten'

(Tompkins and Bird 1972: 4). The positive effects of prayer on plants (Miller 1972) and of music on the growth of plants allude to a possibility of 'cellular consciousness' that is generally relegated to fringe and new age literature, even if this might coincide with the most recent turn within molecular biology, as will be discussed in chapter 7. It can be imagined how making the world sentient opens Pandora's box of efficacy and, more broadly, Pandora's box of 'nature', blurring the separation of an external world to be conquered and controlled by filling it with emotion, culture and sentience.

Beyond plants, making the world sentient also brings new complexities to the assumption, for instance, that all mice would react similarly to a given dosage of *A. afra,* notwithstanding the way they are 'nursed' or taken care of in their cage. The same applies to humans. Generally, it is easier to deny or ignore relations between things and the world in order to pursue the experiment, even if it is quite a task to get these things into the laboratory in the first place as illustrated with the process of making *A. afra* into an object. In the preclinical trial of *A. afra,* the effect of the farmer on the plant is acknowledged and accepted as existing; however, it is not cherished and explored, but is 'dealt with' and minimized as much as possible. This is similar to all of the other entanglements between humans and plants in the world, and it means sentience is not treated as a lawful way of knowing.

Although I am a scientist more than I am a healer, I am less a molecular biologist than I am an anthropologist embedded in life in the 'open air', having spent time with Xhosa *izangoma,* Rastafarian *bossiedoktors* and plants. I am also a mother who uses common plants to heal everyday issues in my family at home, knowing, for instance, that when the little violet flowers grow in spring they are full of vitamins (or life?) to help prevent colds, same as the sap flowing from our maple trees every year. Molecular thought is not foreign to me, since the Canadian context is very pharmaceuticalized and precise chemical compositions can be found on the products I buy. But I know the flowers make us feel reinvigorated. In the latter case, I know they are useful when they make me feel better as well. On the contrary, reading the chemical composition of a product does not give me the impression of knowing anything at all.

As with healers in South Africa, I had learned with healers in Brazilian indigenous Amazonia that plants had different kinds of lives in relation to humans: *plantas sabidas* (wise plants) contained much of the knowledge of the cosmos, while *remedios da terra* (remedies of the earth) were for mundane use (Laplante 2004). The first provide a medium through which to engage in the world if consumed in healing rituals, since they provide visions and help discern (dis)ease. The second serve to heal everyday ills. Both are alive, one to guide healing, the other to sustain life. 'Life' cherished in plants can hence enhance knowledge if one learns how to engage with them accordingly; it is not solely molecular effects that are triggered or oriented in a certain manner, although it can be explained in this manner as well. The plant used in a clinical trial is cultivated in the 'open air', and it is the control of its growth that seems to be contested by the healers. Rastafarian *bossiedoktors* otherwise seem to be playing on both fronts, both indigenous and mercantile, blending how to both tend to a plant in a way that is acceptable to Xhosa healers and, at the same time, suitable to demands of the market. Some Xhosa healers do this as well, tending to permanent 'chemist shops' in the townships. Some plants are sold by stating their high level of iron, for instance, or their use for this or that biomedical disease. While neither buyer nor seller have a sure way to know the chemical composition of a plant or to know disease diagnosis for themselves (which rely upon inaccessible technologies), uncertainty is dealt with through proximity and the felt effects of the plants consumed.

The 'efficacy' and 'safety' concerns emerge as salient for scientists implicated in clinical trials, but embarking on such quality control comes with a restriction to afterwards only trust the commercially prepared and packaged dosage. Safety and efficacy in the townships are dealt with through proximity rather than through distant knowledge of components. While 'chemist shops' and healers borrow some notions from the biomedical domain, such as names of disease or chemicals, they do not necessarily master these; scientists do the same when borrowing some indigenous notions they do not master, such as when they refer to 'trance' or 'possession' in healing and even as they refer to 'indigenous medicine'. 'Indigenous medicine' was overtly referred to in the preclinical trial I followed, although these references clearly lacked an understanding of its ways. Despite the ubiquity and authority of biological understandings of life itself, these are ultimately most meaningful in their original sense to professional specialists, leaving most nonspecialists to doubt their own lived experiences. As such, even the experts usually only 'know' one mechanism, but never all of its possibilities in relating to a body/organism within a medium.

Pollan (2002) wrote a wonderful book, *Botany of Desire,* from the plant's-eye view of the world, which makes one wonder how *A. afra* is related to what is going on in these preclinical trials. In the book Pollan showed how the apple, the tulip, marijuana and the potato coevolved with 'the human desires that link their destinies to our own' (2002: xvi). *A. afra* appears to have made a place for itself in numerous backyards, cultivated fields, markets and laboratories, perhaps for its pungent smell, accessibility, bitter taste, grayish-green colour and furry, sticky texture. Mending cultural as well as natural information, Pollan (2002) discusses the human desires that connect us to these plants as much as the plants themselves, for instance, attending as much to the 'trial and error' that leads animals and humans to figure out which plants are safe to eat and which are forbidden as to the strategies plants figure out to survive in these mediums. In moving through various mediums, *A. afra* is transforming itself as well. Finding a place in the crowded township is a feat in itself, as numerous other plants might have been selected instead. That it has reached global attention is in fact due to the kinship felt with *A. afra* by numerous healers and people showing its utility for healing. This is where Rastafarian *bossiedoktors* and Xhosa *izangoma* share corresponding practices with ways of making medicine as well as where they diverge from the preclinical process from the very onset.

The Rastafarians appeared as important actors in my research, at the interstices between the colonizing scientific trial and the still more mystical Xhosa *izangoma*'s ways of healing with plants. Because of their explicit philosophy in opposition to Babylon and in search of a peaceful, shared future, Rastafarian *bossiedoktors* provide ways of reconciling the new and the old as well as act as peacekeepers between them. The plants are collected in 'wild' areas, dried, and sold in stalls in the markets and during festivals throughout the city of Cape Town and in its periphery. Knowledge of plants is said to come from the KhoiSan, or coloured elders, as well as from daily exchanges with people of all sorts, buying the plants in the markets and reporting their successes and failures, cumulating in a vast array of knowledge about *A. afra* as well as learning from *A. afra* through cohabitation. Although they are philosophically and actively against Babylon, Rastafarian *bossiedoktors,* with their 'commodification' of plants into goods to sell and their adherence to the trade of herbalists, make plants potentially much more amenable to scientific enquiry. In this format it is easier to disband plants from their healing practices and to single out its active molecule. Rastafarian *bossiedoktors* are, however, not recognized as 'indigenous healers', which inhibits their access to permits to collect plants, leaving them in the margins of legalities and legitimacies; however, they feel no threat of their knowledge in plants being taken or transformed in ways they disagree with. The case is rather different with *izangoma,* who more easily become the 'true' holders of indigenous knowledge in current South African politics, even though marginalized in the past, and some of whom have also expressed worry over their secrets being 'taken' by those working in biomedicine.

Rastafarians thus hold a particular space within the Cape. Mostly appreciated for their role as peacekeepers, I have also shown how they prize a particular way of knowing in terms of the *livity* they enact as a way to make way for a world of equality. Words and sounds also partake in this strategy to remain empowered in the struggle with everyday life, and the sacred Herb plays a role in facilitating this path, which is to be done through Jah the Creator and thus with what is given from the earth and its first inhabitants, namely the KhoiSan. In these ways as well as in their aim to heal the nation, their struggle fits particularly well within South Africa's postapartheid 'African Renaissance' and its quest to reestablish the dignity of its people. What emerges within these healing processes is attuned attention to worthwhile ways of living life. In linking this with the preclinical trial's hope that *A. afra* might cure tuberculosis, we here see how these practices tend to the very roots of the disease, which are known to be first and foremost related to poverty and malnutrition. Thus, tending to these issues directly in the everyday through specific foods, worthwhile ways of living with the real and concrete everyday acts avoiding injustice and inequality is certainly a valid route to follow. While in agreement with Xhosa *izangoma* on varying aspects of their practices, namely, in the importance of immersion in sounds, in ways of collecting herbs and in connecting life with broader issues in healing, there are important differences between

Rastafarian *bossiedoktors* and Xhosa *isangoma* practices, to which I attend in the following chapter.

Notes

1. The *Kebra Nagast* is an Ethiopian script, also called *The Glory of Kings*, that contains the foundational myth of Abyssinia. It tells of the Jewish origin and the filiation of their kings with Solomon (Soulimane in the Koran) and Jesus Christ (Issa in the Koran and connected to the Brahmans in India), bringing the holy land to Ethiopia (the new Zion). Written in 1558, it would have been inspired by the Old Testament and the Koran.

2. Two surveys of the health behaviour of samples of older coloureds (sixty-five and older), reported in Ferreira et al. (1996), show how *wilde-als* (*A. afra* in Afrikaans) is the most used plant. The first study (Ferreira 1987) was conducted across an extensive rural area of the southwestern Cape, and a later study (Charlton et al. 1994) was conducted in the Cape Peninsula. What their results indicate is that of seventy-nine people interviewed in the first study and seventy-two interviewed in the second study, thirty-five mentioned *wilde-als* in both cases, with the next most important plant mentioned being *buchu* (*Agathosma betulina*), with twenty-five mentions in the first case and nineteen in the second case. Those were the two most mentioned medicinal herbs. The leaves and stem of *A. afra* are said to be used in an infusion/tea against colds, 'weakness' of the eyes, influenza, diabetes, hypertension, bladder complaints, fever, arthritis, coughs and chest ailments, back pain, as a purgative, as a tonic and as an anthelmintic (Ferreira et al. 1996: 97).

Chapter 6
Dreams, Ancestors and Sound Healing

> Since I accepted the gift of healing from my ancestors, I am
> now a learned or educated person like any other doctor.
>
> – Kubukeli, Xhosa *isangoma*, October 2009

My encounter with a Xhosa *isangoma* was a turning point in my
research; a gesture expressing how *muthi* was to be felt to know it
for healing was particularly telling for my research. In contrast with
Rastafarian *bossiedoktor*'s concrete engagements with some plants
as commodities to sell in their stalls, plants in their materialities are
both less visible and more deeply entangled within *isangoma* practices. In this chapter I explain what I learned with Xhosa *izangoma*.
I first tell of my own experiences with *izangoma* and more broadly
introduce these practices within the current context of the Cape.
Second, I delve into *ngoma* or drumming, which was the way I was
invited to understand the basis of the practice. I then explain the
process of doing *ngoma*, utilizing my observations in particular with
relation to *muthi* and the analyses of Wreford (2008) and, from a
more distant perspective, of V. Turner (1968) and Janzen (1992). I
will attend to *muthi* within these practices, since they are intricately
woven into the fabrics of these entanglements.

Part of the Problem

With Rastafarian *bossiedoktors* my family and I were 'part of the
struggle', but with Xhosa *izangoma*, I initially felt that I was part of
the problem, unable to undertake fully the required investments to

really know for myself. Becoming an *isangoma* is a path that can be followed upon receiving a calling, something that not only needs to be felt but also noticed by someone else. It is thus not a path all can follow. It is, however, a path from which to learn. As with Rastafarian *bossiedoktors,* I first met Xhosa *izangoma* during a meeting at the Indigenous Knowledge Systems (IKS) branch of the South African Medical Research Council (MRC) in November 2007. In particular, I began to get acquainted with Dr. Kubukeli and Mrs. Skaap,[1] who I would both meet again in another meeting in January 2008 as well as visit in their homes and practices in Khayelitsha township in 2009 and 2010. They were two healers implicated in various discussions surrounding paths of reconciliation of their practices within the national context that has opened up to them officially since 2004.

In particular, Kubukeli commented on his disappointment with his involvement with the processes of institutionalization of traditional medicine during our visit to the Delft laboratories' medicinal plant garden, which he had participated in and later dissociated with, as well as his disappointment with the fact that it was *isangoma* trainees who were organizing workshops to 'train traditional healers'. Kubukeli is the *isangoma* who was involved in facilitating Wreford's becoming an *isangoma* with her mentor in Khayelitsha township in 2001, showing his flexibility and open-mindedness in initiating white *izangoma* into the practice, as Wreford (2008) explains in detail. When I met him in his home in January 2008, he was in the process of moving from one home to another after the death of his last wife. The old house was almost empty and the new one almost full with his furniture when I met him, only missing the invitation of the ancestors to the new home. It was once he was settled in his new home, thus, after a few encounters with him, that I asked him how *A. afra* works. This was after visiting his office, where he showed me his certificates in traditional healing as well as some of the exams his students had taken, identifying plants with diseases (as described in chapter 3). Weary of this whole process, I interpreted his 'nonanswer' (or most telling answer), inviting me to a drumming session, as one of being tired of these questions and processes that had little to do with *isangoma* healing. With a clear gesture of his hands dropping down his sides, rolling his eyes (I am not sure if I imagined this or not) and looking towards the drums in the healing room, he explained that one needed to feel the drum sounds to even think of knowing *muthi,* becoming a healer, or understanding anything about the practice. Thus, knowing was to be done 'from the inside', and it was not about obtaining information solely in the mind. This implicit statement was confirmed during the drumming session, as the only thing he said in English, and upon looking in my direction, was that he was 'a literate man, but when I accepted the grandfather's spirits ...', after which he returned to his chanting, dancing and healing initiation performance. The literate and the felt routes to healing moved in opposite directions.

The situation with Mrs. Skaap was very different. Her work was more akin to an *inyanga,* although she was also working closely with ancestors through dreams, namely, with whistle spirits. Her consulting room was filled with dried plants. I visited two other *isangoma* healing rooms; one had only musical instruments, the other a mixture of musical instruments and dried plants. In all four cases, I did not have the chance to witness their healing performances, one of the four being an apprentice *isangoma* working in the Delft laboratories as explained in chapter 4 and it did not seem opportune to impose a visit, Kubukeli being the only one to make such an invitation. It is thus mostly from this very limited experience that I nevertheless speak of Xhosa *isangoma* practices. I will try to offer a broader picture when utilizing my own insights as ways to understand those of other anthropologists, in particular with Wreford's account, which is the most telling, since I did have the occasion to meet her on a few occasions, and her experiences involved Kubukeli and Khayelitsha township. Her account is really 'from the inside' of these practices, while my contribution is only partly so, having mostly learned from the value of such insights to reach closer understandings of ongoing practices. To say that I am part of the problem in understanding Xhosa *isangoma* practices is to acknowledge that to really know them, I would have to undertake such practices over a lifetime. This perhaps applies to all practices of interest in my study, and as such offers a fair positioning. My aim is thus to offer an understanding closest to what I was able to grasp with reference to my objective of finding correspondences between these practices and those unfolding in the preclinical process.

Ngoma

The title of Victor Turner's famous book, *Drums of Affliction* (1968), about the Ndembu of Zambia, is a translation of the indigenous word and concept *ngoma*. Both V. Turner (1968) and Janzen (1992) have attended in detail to the 'drumming', to explain ritualized healing and in an attempt to make *ngoma* more visible as an African institution, respectively. It is also through drumming that I was introduced to *isangoma* practices. I arrived there through my interest in a plant and later more broadly in *muthi*. Throughout I will discuss how *muthi* and *ngoma* enter the picture or tie *isangoma* practices together. Wreford attends to the intricacies surrounding the process of becoming an *isangoma* and all the steps leading to it as she herself experienced them, never isolating drumming or *muthi* from the practices in which they were entangled, making a convincing argument that this should also not be done with 'trance' (2008: 166–71), as there is much more going on. As such, and in doing just that (i.e., isolating trance), Desjarlais (1992) mentions that these experiences of delving into healers' initiations cannot 'represent' a whole practice but only ways of learning, or attuning one's body, to these experiences. As Wreford also suggests, these deepened entries into shamanistic experiences or *isangoma* practices mark a potential bridge between experiences; in Desjarlais's case between American and Himalayan ways of being (1992: 17) and in Wreford's case (2008) between European and African ways of being. I argue that ways of being, of course, greatly differ within these borders, as *isangoma* practices have shown to be highly customized. This is precisely the difficulty in both telling what these practices are and how to take them into consideration within a clinical trial. The solution in the latter case is to mostly disregard them. I will attempt to explain them as I was able to understand them and as they might be recognizable within a clinical process, although that process would need to be transformed thoroughly, as also proposed by a number of scientists involved in the preclinical trial followed in this book. I will not do so as an initiate becoming an *isangoma,* but as an initiate of becoming a more attuned anthropologist, both to these practices and to the accounts of anthropologists who have tried to understand these practices in various ways.

Janzen (1992) provides some of the larger histories of *ngoma* (drums, drumming, doing healing) as studied throughout the African continent. His work helps situate *isangoma* practice in the townships of Cape Town. Janzen worked in the townships of Guguleto, Langa, Nyanga and Crosswords, the last one adjacent to the township of Khayelitsha, where both Wreford (2008) and I met with healers. In the final years of apartheid, Janzen described a scene of civil disorder with police and army repression, a scene in which *izangoma's* role became crucial in providing solace and support to the people in the townships. 'A survey taken in Guguleto found one in four households to be involved with a *ngoma* network; as sufferer-novice, mid-course-novice, or graduated and practicing healer-diviner (igqira-sangoma)' (Janzen 1992: 50). This account matches with Wreford's (2008) experiences as well as my own observations that *izangoma* were omnipresent throughout the townships during the years I journeyed within them. Although South Africa only officially (re-)recognized these 'indigenous' medical practices and began the process of legalizing the practices of traditional healers, including *izangoma,* in 2004, the latter never stopped working, particularly where no other legal instances prevailed, such as in the townships. Wreford's (2008) powerful description of her own initiation in 2001, done before a complex jury of *izangoma,* attests to the still current organization of Xhosa *izangoma* in the Cape's townships.

Janzen attends to *ngoma* more broadly throughout Africa, providing a comparative study of four regional settings: 'Western Bantu, as found in Kinshasa, Zaire; eastern Africa, as found in Dar es Salam, Tanzania; southern Africa, focusing on Mbabane in Swaziland, which is one of the North Nguni-speaking societies; and the townships of Cape Town South Africa, predominately Xhosa, or South Nguni, but also a cosmopolitan synthesis of all of Southern Africa' (1992: 7–8). I focus on the last area, since this is where the preclinical trial I followed took place. Janzen's assessment helps situate this particular context of *isangoma* practices within a larger picture. The Xhosa are part of the South African Nguni migration that came from the region around the African Great Lakes and were well established by the time of the Dutch arrival in the mid-seventeenth century. The Xhosa language, closely related to the Zulu language, is South Africa's second most common home language, with approximately eight million Xhosa people today. 'At the beginning of the 19th century social and political upheavals – known as Mfecane – among the Nguni gave rise to the centralized states of the Zulu, Swazi, Ndebele, and Pedi, and those of the diaspora groups to the north in Zimbabwe, Zambia, Malawi and Tanzania' (ibid.: 36). *Ngoma* was the main way of dealing with adversity, misfortune and sickness, as also seen in the histories of South Africa and of the Zulu kingdom presented in chapter 4. In the period of apartheid from 1948 onwards, these organizations, however, became fragile 'in the context of a divided society, broken homes, deprivation of land and obligation to work in mines, farms of the white man, and in the twentieth century, of the urban settlements and the townships' (ibid.). One can say they moved to the shadows.

According to Janzen, the townships of Cape Town became a place where the various threads of South African society came together in the context of apartheid rule. From his comparative study, Janzen describes *ngoma* in South Africa as far more unitary in its institutional organization than in the other areas, where several dozen functionally specific *ngoma* orders are organized. 'The unitary structure of *ngoma* in Southern Africa combines both divination and therapeutic network building' (ibid.). Thus, there are no named *ngoma* orders, no hierarchy of modes of divining and possession, and '[d]ivination is done without any discernible paraphernalia, more as a Western social worker interviews clients. The third party who "agrees" or "disagrees" with the divination is, however, on hand' (ibid.: 50). In this assessment, Janzen appears to miss out on nuances that need to be made on a number of levels. While I did accompany Kubukeli to the market one day, where he met a women to whom he gave a packet of herbs to help solve some issues she had with her business, the healing sessions I have participated in, as well as those described by Wreford (2008), are nowhere near to ways a Western social worker interviews clients. In the first case, there was no discernible paraphernalia nor was there any third party, while in the second case, both were present and intricately woven together within the healing practices. This difference in ways to attend to issues of different orders may attest to the variety of *isangoma* practices, as healers are relatively autonomous and work with *ngoma* in very customizable manners. However, it does appear as if there are definite more or less sophisticated modes of divining, as well as new more mundane practices closer to those of the herbalist.

Today, *izangoma* can obtain licenses, but only if they position themselves within an association. Such a constraint results in most working 'illegally'. Some *isangoma* trainees are in schools and universities; I met many in the Department of Anthropology at the University of the Western Cape. Others, whom I met at the IKS branch of the MRC, had the task of training 'traditional healers' during workshops. *Izangoma* are increasingly being targeted for both supervision and training by biomedical agencies. 'Between 1999 and 2000 South Africa's Medical Research Council conducted a study regarding the acceptability and effectiveness of traditional healers as supervisors of tuberculosis treatment' (Flint 2008: 190). Healers are also more and more implicated in the fight against HIV/AIDS through various training programs, associations and affiliations, such as their implication with TICIPS' *Sutherlandia frutenscis* trial against HIV/AIDS in Durban as discussed earlier. The Treatment Action Campaign, an advocacy group for affordable medicines for HIV/AIDS patients, also embraces healers in their interventions and has provided workshops for healers since 2005 (Flint 2008: 190). Healers' enthusiasm for these initiatives is lukewarm. While some participate, most are not involved. Wreford (2008) discusses this in detail, critiquing how this is currently being done in a one-way educational manner (from biomedical specialists to *izangoma*) while hoping there is a path towards potential collaboration, suggesting that her role as a white *isangoma* may be strategic in paving the way forwards.

My own observations of this process, as *izangoma* are implicated in the clinical trial of *S. frutescens* led by TICIPS at the Nelson R. Mandela School of Medicine in Durban, show the difficulties in reaching this compromise. The *izangoma* not only have to go through a whole process of literacy, they further need to acquire training in the naturalist ontology, which implies separating mind from matter. For the moment, *izangoma* play a role in maximizing compliance with the trial and have accepted working within the process to deal with HIV/AIDS, but it is difficult to imagine how this process can continue within biomedical meshworks that have set aside all immeasurable ways of healing 'in life', let alone dealing with ancestors, which are central to *isangoma* practices. The Nelson R. Mandela School of Medicine has begun to work with local traditional health practitioners councils to train 375 healers to test patients for HIV, to keep records of their progress, and to refer patients to AIDS clinics where they can receive antiretroviral drugs (Philips 2006). TICIPS organized an *indilinga* (cow exchange ceremony) with Zulu healers in February 2008 to officialize their partnership with healers in the fight against HIV/AIDS within the *S. frutenscis* trial. In 2009, a ceremony took place at the University of the Western Cape to underline the training of healers in tending to HIV/AIDS. Never, it seems, however, are *izangoma* training biomedical health practitioners or scientists, for instance, in feeling drum sounds, unless the latter take initiative in doing so. I suspect this is why Kubukeli appeared discouraged with yet another irrelevant question when I asked him how *A. afra* worked. He had been involved in the process of standardization of indigenous medicine for more twenty years and made several comments indicating his disappointment on numerous levels, which triggered his return to his roots, as he explained. 'Reinventing' traditional medicine within a naturalist ontology does not come without distortion, nor does it necessarily standardize in a useful or 'representative' manner, as I will further attempt to illustrate in explaining what 'doing *ngoma*' implies.

Doing *Ngoma*

Doing *ngoma* is both healing and becoming a healer. It is also both divination and healing. Partaking in these processes are a 'sufferer-novice', a community sharing the affliction, an accomplished healer-diviner, ancestral shades and spirits. Clan ancestors are very important; those of the water, of the land and of the forest, are called upon through various mediums such as animal skins, coloured beads, plants, minerals and *ngoma* songs (Janzen 1992: 96). Connections with ancestors are accomplished through repetitive drum sounds, dances and songs as well as through *ubulawu* (dream foams), which attest to successful communication. It is through these *ubulawu* mixtures that Wreford discusses the place of *muthi*, as explained in chapter 4: namely, a personal combination prepared specifically for the healer and his particular ancestors. Both human and nonhuman actors are hence explicitly part of the process of doing *ngoma*, as nonhuman actors are fully involved in human affairs. In these practices, plants play the role of a medium, within other mediums such as songs, 'which can be activated to transform the negative, disintegrative affliction into positive, integrative wholeness' (Janzen 1992:

191

105). As in Rastafarian practices, words and sounds, as well as intentions, can afflict, transform and heal. The efficacy of the therapy relies upon the combination of these actors connecting, as well as upon techniques and skills acquired through different phases; efficacy is assured, according to Janzen, because 'all in the community feel shared affliction and support the sufferer, even though not all the community is kin' (ibid.: 103).

This kind of efficacy is best explained in Lévi-Strauss's 1949 account of how the shamanic cure is done within the Amazonian context through 'symbolic efficacy'; however, I will introduce some nuances to it that will help in understanding the practices of doing *ngoma* and what I have found to be going on within *isangoma* practices. Of interest in Lévi-Strauss's explanation is his demonstration that shamanistic efficacy 'really works', since it matches mythological and physiological themes, thus binding imagery with real physiological transformations (1949: 20). Csordas (1996) alludes to these efficacies in charismatic healing as well, showing the powers of imaginal performances. Rastafarian *bossiedoktors* told me that a matching of words and acts is necessary to avoid being sick, pointing to the importance of fruitful correspondences between saying or imagining in words and doing. With Xhosa *izangoma*, this entanglement appears to be achieved through movements of opening; bodies, for instance, are opened to loosen them up and provide an entrance for new orders or new ways of being well in-the-world. I will explain this in more detail further on; for the moment it is enough to point to this very real efficacy that occurs through healing performances, with real potential to reconfigure molecular configurations within bodies-in-the-world, as I will discuss in the following chapter.

What is also interesting in Lévi-Strauss's early insights is his proposal to understand psychoanalysis as a modern shamanistic technique, hence linking such practices with those of their predecessors. While I do not agree with this linear understanding of evolution from primitive to modern, I find the analogy enlightening precisely with regard to Lévi-Strauss's argument that myth needs to match with the context to ensure physiological transformation. What is thus lacking in a separation between health practitioners dealing with the mind on one side and other health practitioners dealing with the physiological on the other is precisely that both need to be done together to ensure true therapeutic efficacy. Before I demonstrate how this interweaving is assured through *isangoma* practices, I will present one more nuance. Lévi-Strauss grounds his ideas in a very reductionist form of structuralism, which permits him to grandly simplify practices throughout the world, namely, within subconscious individual or collective lexicons. V. Turner (1968) and Janzen (1992) somewhat pursue this route in describing, respectively, ritual processes and *ngoma* as an institution. My thoughts on preexisting structures are ambivalent. While it is easy to state that in the case of a preclinical trial, there is a predesigned model guiding the way because it is written down on paper and legislated, in the case of doing *ngoma,* it is solely anthropologists who have attempted to formulate such a model and make it appear as if there is a clear path to follow. I argue that there are neither subconscious rules preceding action nor preexisting structures 'out there' in both cases.

Although the RCT model is well-known, it is not a 'thing' in itself, and its complexities, legal and bureaucratic, make it so that none are aware of the same aspects of the model at a given time; even if they were aware of the same design, this information is not known (felt or meaningful) in the same way by all the actors involved, and thus not done the same way. Thus, I follow Ingold and Hallam's (2007) suggestion that we improvise all the time, perhaps even more when we aim to maintain imagined traditions. In this way, while I explain some of the 'phases' in becoming a healer through both Janzen's (1992) and Wreford's (2008) accounts, I tend to bracket these formal steps or core features, assuming they are always being done in different ways. This is most likely the case, since the very process is to reach deepened engagements with ancestors, and is thus open-ended to what these ancestors will tell; the process also takes place in the 'open air', leaving things open to indeterminate circumstances.

Wreford (2008: 165) refers to a great storm climaxing as she underwent the final phase of becoming an *isangoma*, circumstances interpreted as final evidence affirming the success of the process. When I assisted the drumming session, a similar storm, with hail and wind, came to enhance that event as well. The white foam appearing (or not) on top of the *ubulawu* is also left indeterminate, confirming the presence or absence of the ancestors. Many other steps in the process will also be improvised, and a good number of the steps seem to move in a direction aiming to maximize the potential of indeterminacies, thus agreeing with Ingold and Hallam's view that 'there is no script for social and cultural life' (2007: 1); instead, the *izangoma* aim to diminish currently assumed scripts to allow the rewriting of new possibilities. This is done with certain landmarks, although these are also made to appear in different ways.

Janzen describes *ngoma* phases as follows: '1) sickness and therapeutic initiation as a phased rite of passage; 2) divination or identifying the causes of misfortune; 3) associating nosology with "spirit fields"; 4) the "course through the white" of sickness and transition; 5) a sacrifice that sets in motion a circuit of exchanges; 6) the power of the wounded healer, together with fellow sufferers, that is transforming sufferer into healer' (1992: 86–87). All of these acts are 'doing *ngoma*', which corresponds more or less to Wreford's description of her own passage to becoming an *isangoma,* these moments often done simultaneously or during a single session, the 'course through the white' being a lifelong process.

Being sick and entering into a therapeutic process are acts of intermingling with healing in the world in vivo. Going 'through the white' is the actual qualitative transformation of the individual (or group) while entering into a therapeutic process. It is the act of entering into and exiting from the position of the *ngoma* sufferer-novice. Wreford names this the 'sickness calling', in her case manifested through severe depression and moving to Zimbabwe; this state needs to be acknowledged by a qualified *isangoma* (2008: 104). As such, Wreford mentions how in Kubukeli's case, this was done easily by his mother, who was also an *isangoma,* while in other cases it is not always a straightforward process and may take a while for someone to acknowledge ones 'calling'. Once the calling is announced and accepted, specific moments and spatial arrangements are organized to move from one state into another, always involving successful meetings with ancestors and facilitated through diverse mediums, as described above. For Wreford this implied cleansing with her *ubulawu,* used both as emetic and to cleanse the external body (2008: 110). Divination is both to find an explanation of misfortune and/or to begin healing with attempts to act upon the situation. The divination or diagnosis aims at finding out the 'whys', 'whos' and 'wherefores' of an affliction with the help of 'ancestral shades and the spirits beyond'. These spirits are believed to call individuals out of their self-consuming, destructive tendencies (Janzen 1992: 91).

Reaching beyond the mundane for a way out of an impasse is a central technique. Finding the relevant ancestral shades or spirits and the proper mediums to connect through them, including the proper mixture of herbs and bark to make the *ubulawu,* constitute another part of 'doing *ngoma*'.

The 'course through the white' refers to the healing process or transformation as well as the acquisition of the embodied skills to become a healer. This process can last for a great number of years, and I was told it implied seven different steps towards becoming a healer. Janzen mentions that these phases 'in the white' may vary from two to as many as eight, each of which may endure from a few days to many years (ibid.: 101). Many drop out of the process along the way, since it is a lifelong trajectory. The typical process would go from being diagnosed as *twasa* (possessed or called by a spirit), to becoming a novice following the initiation and being coupled with a senior diviner-healer, to moving through the 'course' and becoming a senior, entrusted with aspects of ritual, and finally to becoming a fully qualified *isangoma*. The 'calling' usually manifests itself as an illness, 'epileptic fits, delusions or hysteria' (Crampton 2006: 212). This 'calling' made the afflicted incapable of 'functioning' in the mundane, and this affliction is turned into an expertise in communicating beyond the mundane as a way of divining and healing others within the community. This 'breakdown' is positively reoriented to heal others by intensifying one's connectedness to the world, to ancestors and to others.

Another feature of 'doing *ngoma*' observed in Cape Town and mentioned by Wreford is the sacrifice of a goat (at the time of entry), of a goat or a sheep (at the time of a healer's or novice's purification) and of a cow (at the time of a graduation). Wreford (2008) describes the goat and bull sacrifice at the time of her graduation in detail, telling how the sound the bull makes at a particular moment is read as a sign of the arrival of the ancestors; how the goat's strong will to live after being injured for the apprentice to drink its blood, and its pregnancy, attests to the strength of not one, but of two spirits with which she would be working. Wreford also convincingly tells how the animals are 'treated with respect, blessed before their death, smoked with *imphepho,* and given *ubulawu* to drink in order to facilitate the reconnection with the spirit' (ibid.: 150), a much more glorious reciprocity of the living with the dead than in the slaughter house or even in the lives of laboratory mice, whose death is also named 'sacrifice', this time in the name of scientific knowledge. I have not myself observed such a sacrifice upon participation in the entry of a novice into this process, but I was aware of them as well of goat sacrifices for Xhosa youth entering adulthood. Janzen explains sacrifice or offering 'blood' of an animal as a way to 'restore or regenerate the human community to its ideals' (1992: 104). He further adds that an economic dimension is at work, involving exchange in the distribution of food: 'a communal meal following an all-night communal dance applies tremendous energies to the reconstitution of the social whole that is assembled' (ibid.: 105). Generally speaking, it can be said that most aspects of 'doing *ngoma*' aim to 'turn life around and literally bring life out of death' (ibid.).

Healing through Sound

Turning affliction into power and wholeness is assured through songs. Kubukeli's powerful gesture to explain the importance of embodied abilities acquired through drum beats felt 'in the gut' took me years to assess. It not only turned me towards a whole new literature in the anthropology of the senses, performance and phenomenology, but obliged me to understand all that was behind this way of healing in vivo. Janzen (1992), who made a broader assessment of these practices, mentions some of the more 'formal' aspects of the songs in *ngoma*. He mentions, for instance, how the lyrics of *ngoma* songs from Lemba on the Congo coast reflect the process: 'That which was the sickness, has become the path to the priesthood.' In the Western Cape, the saying goes, 'Let darkness turn into light', reflecting the process of transformation of the harmful into the beneficial. Kubukeli's song in the drumming session I participated in went as follows:

> The healing gift is from our great grandfathers. When I first accepted the healing gifts from my grandfather, people labelled me as someone who is stupid. They even made mockery of me. They did that because they do not understand the calling to healing. Anybody who receives the calling of healing from the ancestors and does not accept the call would experience misfortunes in his or her life. For example, the person may fall sick persistently, may lose his/her job and others. The person may not get well until he accepts the call of healing from the ancestors. It is only through

the acceptance of the call that things will be well with him/her. But since I accepted the gift of healing from my ancestors, I am now a learned or educated person like any other doctor. (personal communication, August 2009)

The song shows the transformative aspect of *ngoma,* which transforms an affliction into a gift, a skill. This ontology of some 'things' being both positive and negative also reflects how *muthi,* or plants more specifically, can be both useful and harmful depending on the skills of the person activating them and for what purpose. This is important for my discussion on *isangoma* knowledge in plants, and is also one of the reasons why the subtitle of this book is 'anthropology in life *and medicine'.* How cosmological, vegetal and animal 'medicines' become 'medicines' entirely depends on the *isangoma's* abilities to activate those medicines in ways that meet specific or general problems with resolve. *Muthi* combinations are also prepared with reference to the specific ancestors with whom the *isangoma* communicates, ensuring such connections are maintained.

'In the Nguni-speaking setting in Southern Africa, the *isangoma,* diviner-healer, is one who (i-) does (sa) *ngoma*' (Janzen 2000: 108). *Ngoma* is focused on the life course, threading and weaving elements of life into a meaningful fabric. Devisch (1993) entitled his book *Weaving the Threads of Life* in explanation of Yaka fertility rituals of southwestern Zaire, sensing these same intensities of rituals. Repetitive drum rhythms of sounds, words and texts performed through songs and dances anchor the affliction in life to change its course. As with Rastafarian practices relying upon specific music, words and movement to be felt as well as lived, the song-dance can be stated to be the basic ontology tying *ngoma* together. It is the transformative power of performed words and sounds that can cause change and make misfortune into fortune. Numerous elements in doing *ngoma* aim to reach a certain state of trance, without reducing the process to this famously described part of healing rituals. The effects of polyrhythmic percussion can facilitate entering certain forms of trance or certain states beyond the mundane, and this is precisely what 'doing *ngoma*' is and how it acts in the life course.

Xhosa *izangoma* expertise is precisely oriented towards maximizing this movement of the opening of the body in-the-environment; by ridding themselves of all forms of standardization, including language, repetitive rhythms, dance, sounds and glossolalia (speaking in tongues) aim to reach a preontological state and bring the patient into this state, as well as provide a means to bring him back into new forms of standardization. In ritual theory, V. Turner explains these as procedures aiming to reach a state of 'homogeneous social matter' so that they can be remade into a new, better, form (1977: 37). Wreford's testimonies tell of numerous events of sheer exhaustion, lack of sleep leading up to states of altered consciousness, and explains the process of becoming an *isangoma* one of 'enduring demands embarrassing or demeaning by turn – kneeling in an *isangoma*'s presence often for long periods of time, lapping up the *ubulawu* like a dog, eating without utensils, going barefoot, remaining celibate, being more or less sequestered, not cutting my hair' (2008: 145), thus pushing some of the limits of humanity to reach higher realms. While she does not mention sounds as central, these certainly took part in the process.

Percussion in the 'metroneme sense' (Chernoff 1979: 49) of the 'offbeat' and the 'hidden beat' pulsates as a basic driving force beneath the surface (Thompson 1983: xiii). 'The supposition that this hidden beat sets up sympathetic echoes with the brain's alpha waves, which are at a comparable rhythm, seems quite plausible' (Janzen 1992: 127), something that might well apply to the Rastafarian *bossiedoktor*'s practices as well. In contemporary practices of *isangoma* healing in South Africa, feeling sounds was shown to me as the central skill to acquire to become a healer. It is the central medium to heal another person. Felt sounds are central in linking with the other senses as well; 'healers, in particular, have a knack for further codification of the senses. A patient's insertion into a healer's particular synaesthetics of turning meaning into matter is part of the healing process' (Stroeken 2008: 467). The mediating capacities of percussion, fundamental to 'doing *ngoma*', reside in percussion's therapeutic vocation. Percussion instruments orchestrate the initiation and facilitate communication with the ancestors who, in turn, provide ways to maintain or modify norms and mundane social prescriptions. Drumming also offers the cement ensuring social cohesion, as each percussionist elaborates a harmonious rhythm. 'The sources of all the texts, dances, and rhythms is this individualized yet collective session in which the participant-sufferer-performer is urged to "come out of his prison" to full self-expression' (Janzen 1992: 129). An initiate, one who can eventually heal properly and provide efficacy through a plant in healing, needs to embody, enact and perform certain sounds and visions through intensive tuning into the world.

Izangoma are involved in rewriting life's story 'in life' in a broad sense. Most of their practices aim to reach preobjective states of being in-the-world, somewhat disrupting current forms of standardization to heal by making way for new orders. *Izangoma* specialize in plant combinations with sounds, song, dance, trance, evocation and enactment, all of which work towards modifications of the complexities of bodily configurations through various manipulations and intensively staged circumstances. The drumming session I attended in Kubukeli's new house, in which he had successfully invited the ancestors into following his recent move, was the first of seven initiating an apprentice to become a healer. More than forty people sat around the room in a tin house in Khayelitsha township near Cape Town, with the healer leading the session, and the apprentice (a woman) in the centre, along with her close relatives and another woman drumming at her side. The opening songs were Christian (later explained as a way to incorporate the influences of colonial missionaries); after the Christian God was acknowledged, the rhythms changed. In these events, there are specific rhythms to praise the ancestors, shifting when it is time to ask the ancestors to connect with the healer-to-be and shifting again when communication is established. This incites the communication pattern of song in which divination is done. Central to connecting with ancestors is *ubulawu* (dream foam), a mixture of medicinal plants and water. A prong-like stick is used to twirl the mixture to form a white froth, demonstrative of a successful connection to the ancestors through dreams; foam on top of the liquid shows that all is going well. Necklaces also acknowledge new connectivity between the healer, the healer-to-be and the ancestors. As the rhythms shift, the woman apprentice and the woman drumming at her side alternate in the centre of the circle, singing, dancing and talking in tongues.

As Devisch describes for healing cults in the Yaka culture, what is being played out is 'a "shape of life" and habitus, that is a tradition and skills that develop and manifest themselves without referring to a script' (1993: 37). The lack of script leaves the possibilities of the performance open. Through patterned rhythms, 'call and response' and narrative performative songs, '[n]goma brings together the disparate elements of an individual's life threads and weaves them into a meaningful fabric' (Janzen 1992: 110). Speaking in tongues (glosso-lalia) similarly calls into question conventions of truth, logic and authority: 'By a semiotic account, then, glossolalia ruptures the world of human meaning, like a wedge forcing an opening in discourse and creating the possibility of creative cultural change, dissolving structures in order to facilitate the emergence of new ones' (Csordas 1990: 24). Glossolalia, together with other practices aiming to 'unstandardize', can also simultaneously create the possibility of creative biological change. Kaptchuk, for instance, suggests 'that ritual healing not only represents changes in affect, self-awareness and self-appraisal of behavioural capacities, but involves modulations of symptoms through neurobiological mechanisms' (2011: 1849). This is achieved through developing enhanced sensitivities in-the-world. It is in reaching this state that a plant can be activated by a healer to 'heal' a 'patient' and have efficacy (or not), practices central in both *inyanga* and *isangoma* healing.

As foreshadowed with Lévi-Strauss's account of symbolic efficacies, we might best understand these healing efficacies as binding mind and body through sounds; whether these be words, songs or rhythms, they need to be felt and imagined in a real manner to bring on desired real transformations. In this picture, *muthi* further enables connections. The *isangoma's* personalized *muthi*, as explained by Wreford, is a connection to the ancestors. When given to initiates, it can facilitate this transformative path towards becoming a healer. In the initiate's drumming session that I witnessed, fresh *umhlonyane* had been set under the woman's bedsheets the night before the performance to purify her dreams. That *umhlonyane* is aromatic might further play into this picture, since, like sounds, scent has a certain physicality (Parkins 2008) that can bind the felt to the 'real'. While some efficacies can be learned by reducing the complexities of the world in a controlled laboratory setting, one can assume that some of these efficacies can also be mastered 'in life', as demonstrated in 'doing *ngoma*'. Hsu explains convincingly that 'recent phenomenologically oriented research points to more intelligent bodies that are primarily practical, that, when they project themselves with intentionality into the world, have the ability and propensity to learn from and test the world in non-random ways' (2010: 57). I hope to have conveyed such an approach and to have convincingly shown how *A. afra* might work in the world through tailored abilities as done by Xhosa *izangoma* and, through different routes and roots, by Rastafarian *bossiedoktors* as well. In this way I also hope to have knit what the world is 'really' made of to its necessary visible, lived and felt realities, enabling consequent efficacies in healing. These efficacies seem to lie in an approach that understands knowing *as* movement, one in which eye and mind are in continuous movement.

Moving Forwards

In this chapter, I aimed to move more deeply into Xhosa *isangoma* practices as I was introduced to them and in conversation with what other anthropologists have been able to grasp. I explained how I found myself to be part of the problem in understanding these practices when I began this research, and how I attuned my anthropological approach to better understand them in practice. I have explained how I understand these practices not as an initiate but as someone who has been transformed by them nevertheless. This journey has led me to tap into sounds and hearing abilities as central in healing, making *muthi* an enhancer and a connector rather than an isolatable commodity containing health. *Ngoma* and 'doing *ngoma*' were explained as tying things together to make efficacious healing through sound. Rastafarian *bossiedoktor* practices pointed in a similar direction, somewhat attending precisely, and perhaps strategically, to the healing routes that are not attended to within imperialist practices today profiling themselves through not so subtle biomedical healing routes linked within global meshworks.

The book began by singling out a plant named *A. afra*, following how it has been tracked down to document its medicinal use and prepared for trial, and reporting preliminary verdicts on its potential use against a tuberculosis pandemic. While this objective of the preclinical trial was fully attended to by scientists and agencies working towards the clinical trial, in the last five chapters I have instead attended to the preclinical trial's second objective, which was to 'recognize indigenous medicine'. The following chapter aims to explain the ways molecular biologists partly weaved these apparently incommensurable objectives together within their practices as well as within the preclinical process of testing *A. afra* against a tuberculosis pandemic.

Notes

1. The first *isangoma* presented himself as an *isangoma* doctor, as indicated in the epigraph of this chapter, while the second names herself Mrs. Skaap on her healer's business card. I will refer to them respectively as Kubukeli and Mrs. Skaap in the rest of the chapter.

Chapter 7
Weaving Molecules with Life

At this molecular level, that is to say, life itself has become open to politics.

– Nikolas Rose, *The Politics of Life Itself*

Molecular biologists paved my entry into the preclinical process, sharing their concerns with the notion of efficacy in dealings with indigenous medicine. Concerns with efficacy within the preclinical process at the interstices between a quest to find a cure against *Mycobacterium tuberculosis* and a quest to retrieve African dignity are played out around life, modifying, controlling, anatomizing and engineering it at the molecular level. In this chapter I show the ways in which molecular legitimation is dealt with by the molecular biologists involved in the preclinical trial of *A. afra*. First I delve into the science of molecular biology, and how it has opened towards new emerging forms of life in the last half century and increasingly so during the last decade. I explain how the creation of new forms of molecular life may point to commensurabilities with ways indigenous healers rewrite life's story. Second, I show how this potential commensurability between biomolecular practices and indigenous ways of reorienting life was not fully taken up by molecular biologists involved in the preclinical trial of *A. afra*, rather opting to follow a more classic agenda in molecular biology in its dealings with indigenous medicine. Third, I explain how when a full step forwards is taken towards intervening upon life as a movement of opening, it is done under the banner of scientific innovation, letting the indigenous issue wither away. A final section discusses the push and pull between indigeneity and scientific innovation set in opposition to one another, an opposition that can be bypassed by thinking in terms of improvisation. Through these discussions I hope to weave together the visible, the lived and the felt as inextricably entangled with the real in envisioning what kind of world we have in common. This is done in part by the actors themselves. As such, the practices of the molecular biologists, like those of the healers, are not fully in line with the predesigned clinical process.

Life in Molecules

Molecular biology took root at the Rockefeller Foundation in the United States during the years 1932–59, pouring about $25 million into the molecular biology program, more than one-fourth of the foundation's total spending for the biological sciences outside of medicine (Kay 1993: 6). Warren Weaver, the director of the Rockefeller Foundation, coined the term 'molecular biology' in 1938. Borrowing methods from physics, mathematics and chemistry, the new biology also borrowed from other life sciences such as genetics, embryology, physiology, immunology and microbiology, with a promise to uncover the 'secrets of life'. The new biology stressed the unity of life phenomena common to all organisms, studying fundamental vital phenomena at their molecular level, such as in bacteria and viruses. Aiming to discover general physiochemical laws governing vital phenomena, 'it distanced its concerns from emergent properties, from interactive processes occurring within higher organisms, between organisms (e.g., symbiosis), and between organisms and their environments, thus bracketing out of biological discourse a broad range of phenomena generally subsumed under the term "life"' (ibid.: 5).

By defining life in terms of fundamental physicochemical mechanisms, molecular biology ultimately narrowed its principle focus to macromolecules and macromolecular mechanisms found in living things. Mechanisms are studied to understand, predict and control life at the molecular level. The project of biology, to understand life in plants and animals, progressively led in molecular biology to the 'disappearance of naturally occurring plants and organisms from the field of molecular biology and their replacement with new entities; and, second, it meant the culturing of these entities in the laboratory' (Knorr-Cetina 1999: 140). I have shown in the first chapter how *A. afra* was mostly disentangled from practice and mediums as it was reported and documented in botanical inventories. 'Life itself',[1] at its molecular level, has also been progressively extrapolated from its environments to become an object upon which to intervene. While initially arising with only indirect links to medicine, it today occupies centre stage in the making of biopharmaceuticals cultured in the laboratory to resolve world health issues. Bringing a naturally occurring plant into the laboratory is thus a first great challenge faced by the molecular biologists implicated in TICIPS, since doing so fell outside of the dominant practice. It is through the 'complementary and alternative medicine' branch of the NIH that it thus appears. A second great challenge is to bring indigenous knowledge into this very narrow picture. The recent turn in molecular biology gives the impression of being able to do so, although arduously and within restrictive measures, as I will discuss in the following sections. First, I will overview early molecular biology and its more recent turn.

The rise of molecular biology really began in 1953 with Watson and Crick's 'discovery' of the double helical structure of DNA, well described by Latour and Woolgar (1988).[2] Following this 'discovery', the central dogma in early molecular biology became 'that DNA (deoxyribonucleic acid) contains the building blocks of life' (Knorr-Cetina 1999: 139). Simply, the dogma states that 'DNA makes RNA (ribonucleic acid), RNA makes protein, and proteins do almost all the real work of biology' (Gibbs 2003: 48). Through increasingly sophisticated visualization techniques, it was found that information was stored in the twisted ladders of DNA, more specifically, in the chemical bases of four kinds of molecules, A, T, C and G (adenosine, thymine, cytosine and guanine), which pair up to form the rungs of the ladder. Living cells use the four basic molecules each composed of atoms named nucleotides. The alignments of these nucleotides in a precise order are genes. 'The dogma holds that genes express themselves as proteins' (ibid.: 49), gene expression thus becoming 'life itself'. This implies that the ordered variety of life, which biology had investigated before, is really an endless variation of the same. 'Classical biophysical, biochemical, and genetic techniques can all be seen as aiming at the construction of an experimental environment in which it is possible to replace the milieu of the living cell in such a way that, starting with "model" organisms, cellular structures and/or metabolic processes can be isolated and analyzed' (Rheinberger 1995: 251). DNA, of course, exists nowhere on its own, yet somehow an idea emerged that by understanding its mechanisms in a controlled laboratory environment, one could produce the 'book of life'.

The Human Genome Project was announced as complete in early 2001, following the dogma that had, however, set aside 98 per cent of the DNA, labelling it 'junk' DNA (Lock 2005: S47). It only took a couple of years until this 98 per cent was found to produce RNA implicated in gene expression and regulation, and hence implicated in the multiplicities of ways for bodies to come into being in environments. The basic structure was therefore not enough to 'determine' the organism. With 'junk' DNA back in the picture, the unfolding of life opens to infinite possibilities. While medical intervention had been restricted to the level of metabolic performance, the advent of recombinant DNA technologies in the 1970s opened up the possibilities of instructing metabolic processes (Rheinberger 1995: 252). Since DNA recombinants are themselves macromolecules, molecular biology has shifted to interests in understanding how DNA recombinants interfere with the organism as a whole, since they participate in virtually every process within cells. 'Consequently, the research interests of molecular biology are no longer confined largely to mapping structure but have expanded to unravelling the mechanisms of cell and organ function through time' (Lock 2005: S47). To understand the 'essence' of life by reading and reporting the 'laws' of nature in older forms of biology thus became, in molecular biology, the overwhelming complexity of a myriad number of well-regulated biochemical steps.

'From now on it is no longer the extracellular representation of intracellular processes – i.e., the "understanding" of life – that matters, but rather the intracellular representation of an extracellular project – i.e., the deliberate "rewriting" of life' (Rheinberger 1995: 252–53). The space between a desire to understand life and one of acting deliberately upon life-making processes at the molecular level is where the preclinical trial of *A. afra* is being played out in the laboratory as well as within the politics these imply. Understanding life in the laboratory is a modern project, one within which the RCT is designed. Its results guide ways to administrate populations, in line with Foucault's (1976) explanations of these biopolitics, also discussed in chapter 4.[3] Deliberately 'rewriting' life in the laboratory corresponds to recent developments in molecular biology. As this new research agenda is still organized within the RCT model, its results are also destined to provide ways of administrating populations, yet because they no longer aim to simply 'understand' or 'represent' life but rather aim to 'reorient' or 'rewrite' life at the molecular level, they become more visibly open to politics.

In this recent move, molecular biology in the preclinical trial of *A. afra* becomes more open to indigenous medicine on three levels. First, in its materiality, since the project is looking for molecular configurations in a plant used in practice, rather than in a synthesized molecule (even if I have shown that *izangoma* do not rely heavily upon the materiality of the plant in healing); second, in its very project, since *izangoma* also deliberately aim to rewrite life's story; third, in its politics, because one of its aims is to address issues of indigenous dignity within the South African context and beyond (which may on this level conflict with a humanitarian horizon). On all three levels there are, however, important nuances that can all be explained by a difference between doing medicine in a controlled environment as opposed to doing medicine in the midst of life-making processes. As such, molecular biologists involved in testing an indigenous medicine pull interests in 'life itself' towards recreating the conditions under which the molecules and reactions occur in the organism (the plant) as well as in their engagements with healers; however, they do so only very shyly, rather rapidly closing this opening to construct a new object through varying technological procedures.[4]

A more attuned interest in metabolic performance has become one of reprograming metabolic actions, not solely interfering with them. Rose calls this the emergence of new forms of life because it is knowledge that opens 'ways in which life may be mobilized, controlled and combined into processes that previously did not exist' (Rose 2007: 15). In a phenomenological perspective, this is and has always been the case, and hence this is not necessarily new. What is perhaps new is the renewed attention to these possibilities within the laboratory. Rather than imagining solely understanding and restoring life to its 'intact' or 'normal' state in a movement backwards, the work is thus imagined as a movement forwards orienting 'life itself' as well as acting beyond this scope, very much in line with what I propose in a phenomenological approach in anthropology as well as with what *izangoma* do 'in life'. In molecular biology, narrowing life down to physiochemical processes enables reifying 'life itself' as an object upon which to intervene. Enhancing life's performance is, however, also narrowed down, generally modifying it with a found molecule 'out there' that has been understood upon a temporarily passive body-object, thus screening out attention to numerous possible processes that can come into play and that have different political horizons. In the preclinical trial of *A. afra,* politics of life itself played out at the molecular level in the laboratory have immediate preoccupations with issues of indigeneity, somewhat imbued with what the *izangoma* do 'in life' in its broader sense and thus clouding the horizon.

Since Foucault's assessment of modern forms of biopower, many anthropologists have offered explanations as to how this investment in life may have transformed itself with the turn of the century, as briefly touched upon in chapter 4. For Fassin (2000), these biopolitics have opened to politics of the living on one side and to politics of life on the other; the former politics would play themselves out in the research laboratory, while the latter have refugee camps and humanitarian rights as an ultimate horizon. Anthropologists studying politics of the living have distinguished the emergence of new forms of biopolitics within the Human Genome Project. For Rabinow (1996), while biopolitics would administrate populations on the basis of knowledge gained in biology and from a process of discovery of the 'laws of nature', as in early biology, current forms of biosocialities would orient biological research according to perceived social desires. Biosocialities imply a looping reflexive effect, making it so that social desires feed back into ways of orienting biological research. In this paradigmatic shift, life at the biological level becomes malleable, although still working in concert with disciplinary rationalities.

The preparation of an indigenous plant for investigation in an RCT can appear to embed an overlapping of the two forms of rationalities, disciplinary and postdisciplinary, as on the one hand it answers to appeals of indigeneity, but on the other hand, it completely shuts this conversation out. The experiment with *A. afra* at times relies upon a model designed to reveal the 'laws of nature' (the plant's molecular configuration) when it aims to recognize indigenous medicine but at other times relies on newly made forms of life (transforming the plant into animal feed, for instance), in both cases consistently disregarding engagements between organisms in the environment. Further, disciplinary rationalities in the South African colonial context were less based on knowledge upon individual bodies and more done upon the 'African' population, as noted by Vaughan (1991). Postdisciplinary rationalities thus appear enmeshed in the setting apart of a whole people through politics of indigeneity, with an elite sometimes bringing indigeneity into the 'traditional' and fixed in the past rather than making it appear as innovative ways of orienting bodies within mediums.

Haraway (1993) for her part rather sees biopolitics of postmodern bodies completely superseding disciplinary rationalities, a claim supported by Rose (2007). The argument that postdisciplinary rationalities are unilaterally in effect, however, disregards two things. One, it disregards the fact that it is only a very small elite whose desires may orient biotechnological innovations. Second, it disregards the fact that there are more than solely disciplinary and postdisciplinary rationalities playing themselves out in the world, as shown in the South African case. More generally, making the possibilities of reorienting life the sole domain of biotechnologies dismisses the possibilities of doing so 'in life' through bodily cultivation, attuned attention, technologies of the self and deepened engagements in the world, as done by healers. All of the scientists I worked with are highly aware of these limitations and all mentioned that they aspire to learn from indigenous medicine in life, but they have not found the means to do so within their science. As such, life in molecules, resting upon a narrow notion of 'life itself' will move beyond laboratories and pour into politics of life at the heart of the humanitarian rationale (including the military): human rights, in the name of which emergency action is proclaimed (such as in declaring a pandemic), are rights to 'bare life' before they are citizen rights (Fassin 2000).

To avoid a biopolitical reading of the politics of life because of the hierarchy of lives it supposes and perhaps even more because of the disappearance of subjects it cultivates, Fassin proposes to take into account both the basic 'human rights' to live as well as the human rights to have a life worth living (2010: 88); 'to live beyond and to keep on living, at the frontier of the biographical and the biological' (Derrida, in ibid.: 84). He shows how in the South African context of the HIV/AIDS pandemic, more than 'life itself' is at play in what he calls the ethics of survival. He follows the biographies of three South Africans to bring further subtlety to this notion of survival, which entangles biological and lived experiences. From those biographies, Fassin points out how through 'humanitarian' interventions upon life itself, or through antiretroviral medication and CD4 counts (a measure of immune activity), other forms of life are indirectly attended to; to keep the body alive, shelter, food and funding are provided, and people remake a life larger than its biology. Fassin is left satisfied that even if the richness of life is only indirectly attended to in humanitarian interventions, such interventions provide ways for people to do so themselves. This, however, does not seem to resolve the issue of the hierarchy of lives. In fact, I see no reason why it would be more harmful for anthropologists to provide a biopolitical reading of life than it is for biologists to do so, as Fassin argues (ibid.). In both cases, it seems that it is the entanglements in life-making processes that need direct attention.

Fassin's argument that we 'need to return to an inquiry about life in its multiple forms but also in its everyday expression of the human' (2010: 94) is thus combined with the need to attend to the multiplicity of biologies. The notion of 'local biologies' to account for biological and social life as mutually constitutive proposed by Lock and Nguyen (2010: 1), can be a reminder that human bodies are not everywhere the same, precisely because bodies are in motion and in constant emergence within contexts. What is, however, not accomplished with the notion of local biologies and with the idea of an ethics of survival is the dislodging of empirical truth in 'life itself' as primary; in this way, lived and felt experience remains secondary. In both cases 'life itself' is not fully problematized. Life in molecules attends strictly to orienting cellular life, namely, through enhancing immunity, as in our case study. Through this route a better life or dignity for a people is hopefully provided, but without attending directly to the latter issues or to abilities to do so in the midst of life-making processes by attuning to the medium, as is done by healers. Reorienting life in molecules clearly plays into life in its broader sense, yet in a blind way, as in an experiment. In the preclinical trial of *A. afra*, possibilities of reorienting life at the molecular level in the confinement of the laboratory are further limited in its abilities to hear sounds emerging in life.

Sounds in Life

Survival, or the struggle to remain alive or living, appears throughout the *A. afra* trails and trials I followed. Rastafarian *bossiedoktors* pursue a struggle to simply live, and both *izangoma* and molecular biologists attend to a process of 'rewriting' life's story with objectives of extending or meliorating life beyond molecules. Survival as a real life-making process in mediums breaks with an idea of 'life itself'. A phenomenological stance remains in the immediacy of emerging engagements in life. It is in this continuous motion that we stay alive, always in mediums, with hopes and fears, feelings and thoughts, eye and mind entangled. It is in this continuous movement of opening in life that the preontological is important to the phenomenological project. It is the roots of experience, making the world appear in new forms through never-ending engagements.

People, plants, animals and any living being are always concerned with survival, whether it be for one's own immediate survival or those of the nation and/or world. It is never possible to completely stray away from a preontological stance. Yet subject matter can become covered up, can deteriorate or become hardened, making itself more difficult to be grasped; indigenous medicine can, for instance, become veiled and utopian, it can be instrumentalized to claim rights or legitimacy, while RCTs can make medicine into precise molecules, commodities assumed to contain health. These molecules are shaped into hardened healing objects through legal entities as well as bureaucracies. These complexities make it challenging for a phenomenological account as well as for *izangoma*; however, they are perhaps vital in a broader sense. Ridding ourselves of accumulated forms of standardization in the everyday to make place for new orders is necessary to ensure meaningful engagements in making a common world. 'Essentially the person exists only in the performance of intentional acts, and is therefore essentially not an object' (Heidegger 1962: 73). We cannot, therefore, objectify the person to reach an understanding of being, nor can we objectify 'life itself' and reach such an understanding, let alone 'rewrite' or heal it in a meaningful way. It is rather in the performance of acts through which some understandings and healing might be found.

Simply living can be relegated to secondary status and dealt with at the molecular level when threatened on a population level for any reason. In the reverse, this state can be cherished as the highest state of consciousness in places in which people live and die; in dealings with life, hope and death, performances aim to reach near-death experiences in a quest to further know vibrancy in life. The preontological state can be aesthetically embellished and fine-tuned involving the heart and soul, or it can be simply maintained, kept 'intact' through procedure and clinical routine, hoping meaning will reappear in attending to physiochemical mechanisms. With my intellectual interests regarding A. *afra,* I received polite answers and collaboration. When my husband accompanied me to the townships and played music with the Rastafarian family, we received admiration and respect and were accorded value and wisdom. A Rastafarian *bossiedoktor* made this point clear during one of our conversations, showing in particular how it was my husband's ability to attune to sounds that provided him this wisdom, while information acquired intellectually was interesting but not necessarily meaningful.

Living in the Cape throughout this research, I was reminded of my survival on a daily basis. Our house was broken into upon our first week of arrival, we received constant warnings to not walk alone in the hills and mountains for fear of muggers, and violent winds and storms, reminding us of the strengths of 'nature', were both vitalizing as well as threatening. People living in extreme conditions, dying of disease, gunshots, car accidents and so forth, were a daily affair. Ross (2010) describes life in the townships of Cape Town as 'raw life' in the sense of people 'surviving' through the everyday. I imagine this supposes life elsewhere to be 'cooked' or more refined or perhaps filled with other mundane occupations. Of all the places I moved through in my study, it was in the townships of the Cape where I felt the warmest sense of humanity, or perhaps the most thorough and meaningful engagements in the world. Living in the Cape gave an edge to the everyday, fulfilling life with extreme senses of humanity and extreme feelings of being 'together with' others in life.

This urgency and vitality transcended the positioning of some of the actors partaking in the preclinical trial of A. *afra.* It was both a quest to solve a world pandemic and a more immediate crusade to retrieve the dignity of a people, their own or those of a nation and world. In these respective hopes, forms of engagement within the preclinical trial, survival was externalized in order to make way for molecules. In doing so the trial barely skimmed the surface of the indigenous medicine from which it aimed to innovate biopharmaceutically. The notion of 'life itself' is central to the life sciences, yet the life sciences are utterly detached from what makes a person alive. 'Being alive' is precisely what seems to be left to wither away when dealing with 'life itself' in an RCT, even as it opens to the medium in its molecularization, as only 'vital' chemical processes are dealt with. Yet, life is also maintained through a desire to live. It is the condition of being alive and wanting to be alive in the world that both humans and nonhumans share. According to Ingold (2011), we need to return to a preontological stance; we need to get back into the river, to the flow and fluidity of the movements in life, rather than remain on the banks, where life is much less dangerous. This riskier route is the one that has to be followed in order to understand, connect and heal through ways of engaging in the world. It is a return to a ground by no means solid, yet always moving, carrying on, with no beginning and no end, simply flowing with all its might, with high and low tides, 'scourging the banks on each side and picking up speed in the middle' (Deleuze and Guattari 1980: 359–60). 'We have, in effect, been concentrating on the banks while losing sight of the river' (Ingold 2011: 14), even in anthropology.

These ways of knowing and not knowing life I have proposed to understand as more or less attuned attention to simply living. As such, interests in 'life itself' do not want to know life as it is lived. Rather, it is a concern with molecular life that in the end exists nowhere else than in the RCT experiment, and even rarely so within these environments. Life in molecules is in this way extrapolated from the world, aiming to return to the world to enhance life's possibilities. It appears to be known as well as cover all forms of life; however, indeterminacies abound as well as remain confined to narrow mechanisms that exist nowhere on their own. The immune system as life itself is similarly stated as 'everywhere and nowhere', but located in the human body as well. It is also to be 'known' in test tubes and enclosed cellular reactions. These performances in the RCT are, however, done in life. They enact politics and histories of a colonial past aiming to reconcile indigenous dignity within a global world.

Life in molecules can partially blur the boundaries between nature and culture as well as between humans and nonhumans or between different forms of life; 'genome mapping showed that human DNA is closer to that of other living organisms than had been anticipated—we share more than 98 per cent of our genes with chimpanzees and about 35 per cent with daffodils' (Lock 2005: S49). Perhaps we share as many with *A. afra,* and this could in part explain connections with plants as demonstrated by *izangoma.* What biochemists study in *A. afra* are, however, its molecules of 'life' (molecular configurations, bioactive compounds, biomarkers, active principles) and how those molecules can encourage human life to inhibit disease. While the ways of transforming *A. afra* in the laboratory are definitely new and experimental and not a representation of the plant, the process of the preclinical trial of *A. afra* did present itself with an intention of recreating some of the conditions under which the plant itself can show efficacy in the organism.

In the first moment of TICIPS's emergence in 2005, it was hoped that the trial would simply attest to the veracity of ancestral roots; thus, TICIPS tested the whole plant rather than an isolated molecule, attempting to mimic indigenous practice and its synergies. This was thus done in a project to understand life in *A. afra* in a way that would provide recognition of indigenous medicine. Pursuing more 'innovative' routes in molecular biology, which points towards more commensurabilities with skilled *izangoma* also working towards 'rewriting' life's stories, would have to involve skilled healers in the scientific process. *Izangoma* are involved in rewriting life's story 'in life' or through the medium without the intermediary stage of representing its process 'in vitro'. They do not restrict life to molecules that can be invested, controlled and manipulated. Further, the person to be healed is widely expected to participate actively in the diagnosis and cure, as opposed to being acted upon. *Isangoma* practices take as an ontological starting point the condition of a human beings' engagement within life-making processes. Practices are enacted in the 'open air', investing life through and through in the medium; the tin houses where healing events are enacted have numerous openings, the floors are earth, and they are situated in and involve the neighbourhood where people live and die. *Izangoma* specialize in combinations of plants used with sounds, song, dance, trance, evocation and enactment, all of which work towards modifications of the complexities of bodily configurations through various manipulations and intensively staged circumstances. Life in plants, and in *muthi* in general, purify as well as connect *izangoma* with the ancestors in a movement forward linking pasts with futures. It is life that can be activated at a distance. It is life that is never restricted to its objective materiality.

As such, healers claim that the *A. afra* cultivated for the RCT has lost its 'life' or efficacy; it has lost its meaningful entanglements with people in mediums. TICIPS's South African director recalled one healer walking through a laboratory who, upon seeing a plant in a pot, remarked that the plant was 'choking'. I have seen, tasted, felt and smelled *A. afra* growing in the yards of many healers, cultivated or tended to as well, but there are ways of collecting, preparing and cohabitating with the plants that are not of control or mastery. While *izangoma* are not constrained by the RCT per se, since the process essentially dismisses them as lawful experts, the practices of the *izangoma* at the edges of the preclinical trial helped me understand how making a medicine work in the inhabited world can be accomplished. In this way, it is not explicitly through manipulating molecules that the whole organism is invoked by the *izangoma*, yet it is through attuned engagement in the world that this can be done, as the healer brings another person to be healed into his particular synaesthesia.

Isangoma healing practices rely upon acquired bodily skills to modify bodily configurations within the inhabited world; healing is more about turning meaning into matter than relying upon 'matter itself'. The healing process thus end in 'objects' rather then begins with them. To the contrary, molecular biologists work to make 'objects' that will later be used, if successful, to heal in themselves. Their insights into microworlds of molecules provide thorough representations into potential specific intra- and extracellular processes that can modify bodies (or not); however, they do not attempt to do so through research in *A. afra,* instead aiming to show how its molecular configurations can attest to indigenous medicine as if the latter were fixed in time. Healers are, however, not confined to the molecule per se as the way of knowing how to invest life, nor do they know, or want to know, about its particular workings in terms of mechanism or representation of something 'already there'; they 'know' through attuned attention in life, which is open-ended and always in the making.

The *isangoma* ways of knowing or letting me 'know' how medicines may heal have been by way of inviting me to attune my attention in these matters. They required me to 'feel' with them to understand what they were achieving. Although molecular biologists I met during my research explained that their experience and intuition was crucial in decision making, these biographical dimensions of embodied work remain black boxed and absent from research results. They also did not guide me to share their intuitions through experiencing the molecules. Scientific enquiry has moved away from feelings or engagement of the senses since its earliest forms of experimentation. According to Knorr-Cetina, 'two factors aided in the 'disembodiment' of science; one is the inclusion, into research, of technical instruments that outperformed, and replaced, sensory bodily functions. The other is the derogatory attitude important scientists developed towards the sensory body' (1999: 94). While the molecular biologist's body remains a precondition to conduct experiments (Merleau-Ponty 1945), sensory skills do not constitute the fundamental conditions, even if they do participate in scientific enquiry: 'If anything is indeed irrelevant to the conduct of research in molecular biology, it is the sensory body as a primary research tool' (Knorr-Cetina 1999: 95).

Knorr-Cetina (1999) distinguishes between the acting body, the sensory body and the experienced body to nevertheless show the extent to which scientists remain embodied. The acting body is involved in scientific research, since experimental work is manual work; however, 'the scientist's body as an information-processing tool is a black-boxed instrument' thought to work best 'when it is a silent part of the empirical machinery of research' (ibid.: 99). A particular sense of vision remains essential to this paradigm. The rest of the body, especially the hands (usually gloved), are made to work for the eye in laboratory work, remaining merely accessories: 'the chair enables sitters to think without involving the feet at all' (Ingold 2011: 39). The thinking performance of the biochemist remains of prime importance in establishing the difference between the medicine 'working' (or not) according to the desired verification process. The biochemists may or may not succeed in finding the relevant molecular structure within the plant at that particular moment, or they may or may not succeed in determining a particular relation with the disease the plant is being tested for. Intuition, tacit knowledge or experiential knowledge can decisively shape success or failure in designing the relevant model. These skills are, however, not part of the explicit requirements in preparation for the RCT; rather, they are asked to disappear. The intensity of bodily expertise developed by the healers is astounding, yet, as with the biochemist, it is not part of RCT legitimacies, given the resistance of bodily experience to causal logic. This exclusion of bodily expertise, or its denial, takes most of the *isangoma*'s legitimacy away from the preclinical process, since it is precisely in that expertise that *izangoma* excel.

As for the biochemists, preparing for the RCT allows them to still represent their knowledge and maintain, even gain, their legitimacy. This occurs through a reification of the 'medicine itself' as an object of healing or of life reorientation. In life embodied, *umhlonyane* fits into this picture as a discrete medium through which one can become entangled in more or less useful ways. It is less the life in plants or in their molecules that is useful then the ways the plants are embedded in life as well as the relationships established with them. Cultivated fields of *A. afra* are associated with colonial invasion of the land by Afrikaner farmers, hence, not a neutral ground. 'Biomedicine was a means of colonial domination in South Africa and explicitly so since 1948, marking the beginning of the apartheid era' (Flint 2008: ix). It marked the decline of indigenous healing and spurred a struggle for legal recognition. The RCT is a legislative research design that continues to marginalize these indigenous ways of knowing life and medicine, even with increasing apparent commensurability with highly prized biomolecular knowledge in a notion of life as a movement of opening. Divergence between ways of knowing life in molecular biology and in *isangoma* healing practices relies on possibilities of representation (only one practice is the current legal route), as well as in possibilities of distancing oneself from the knowledge produced to enable the made medicine to travel to heal universal biological bodies.

Bodies in the preclinical process are made into objects; they, like the plants, are not only homogenized in their biology, but they are almost completely disconnected from their environment. Each and every process of the preclinical process is about dissociation from the world. While it aims to act 'objectively' on life itself (strictly speaking, on molecular life), the preparation for an RCT process does so by disbanding the elements that may be giving the body its very life. The molecular biologist finds a precise way to prepare the plant that dissociates it from its medium; more importantly, the molecular biologist can dissociate his knowledge from his own body, even if it is through his body that he has gained this knowledge and even if it is for his body, or for all human bodies, that the molecule or plant preparation is designed (as all bodies are assumed to be made the same). Life is reduced to one active principle, or to one molecular configuration or preparation that can maintain vital necessities. As Ingold (2000: 382) notes with regard to DNA, this molecule, however, never existed on its own, except when artificially isolated in the laboratory. It is only when the selected/made molecule or preparation enters the living machinery of the cell that it has the effects envisioned; hence, it depends upon the total organismic context in which it is situated to (re)act.

The RCT design, however, predicts that a molecule will react in the same way in all bodies notwithstanding the medium in which this body thrives. It seems the ontological starting point, the condition of human beings' engagement in the world, is sidelined for the time being. It would otherwise make the dichotomy between nature and culture obsolete. If the new emergent forms of life are ones in which biology is no longer destiny, as discussed by Rose (2007), in this undoing it has made human manipulation of nature even more important while still leaving the relations between things, bodies, plants and people in the world mostly unattended. While engagements in-the-world are more difficult to represent and control, these interconnectivities might correspond more closely to how social ties emerge as well as to how uncertainties and indeterminacies in life are continuously dealt with. It might also permit attending to the dignity of a people. Rather than opening to the medium in which people and plants live to grasp how these engagements play into gene expression, and rather than turning towards indigenous medicine to learn from these engagements, the recent turn in molecular biology is only set into motion once it lets go of the assumed confinement to the past linked with indigeneity.

From Indigenous Knowledge to Innovation

To move beyond a restrictive project of translation of indigenous medicine into a scientific understanding, TICIPS was rebaptized The International Center for Innovation Partnerships in Science in 2010. In this way the preclinical trial I followed moved very much as the South African 1932 Code of Native Law did, enabling 'innovation among the white medical practitioners while denying it to Africans' (Flint 2008: 4), showing a clear monopoly in moving forwards only for selected practices, leaving other practices at the edges or to be co-opted under scientific profiles. In its opposition to 'science', the 'indigenous' appears to refrain from being part of making innovative therapies. Even in stating that 'indigenous knowledge systems were way ahead of the popular idea that the drugs of the future would be based on our DNA, very personalised and specific to the needs of the individuals' (Johnson 2011: 1), moving forwards in scientific innovation (or more precisely into the bioeconomy) implies moving beyond indigeneity. It will be done through new forms of medicine at the cutting edge of biopharmaceutical, bioinformatics, biomolecular, biotechnological and biomedical innovation, under the scientific banner, with human rights relegated to the edges.

Molecular biologists' dealings with indigenous medicine appear as a way to legitimize ancient roots, yet without embracing these as lively and anchored in sophisticated engagements between people and plants. Knowing there is molecular variability in the plant depending on its setting and its manipulations by healers, or any plant caretaker, the biochemists involved in the preclinical trial I followed sought to minimize those variations rather than learn from them. Preparation for the RCT requires genetically identical plants in order to facilitate finding and reproducing an active ingredient of the plant, an ingredient that can then be 'commoditized, transported between laboratories and re-engineered by molecular manipulation, the properties of the plants transformed, their ties to a particular individual living organism, type, or species suppressed or removed' (Rose 2007: 15). Molecular biologists are well aware of the multiple potentials found in a plant depending on the ways it is prepared, and they know the RCT is restrained in exploring this matter.

In a comparison between laboratories of particle physics with those of molecular biology, Knorr-Cetina tells us how 'in the molecular biology laboratories studied, in contrast, the phenomena assert themselves as independent beings and inscribe themselves in scientists' feelings and experiences' (1999: 79). She further mentions how molecular biology laboratories were based on interaction with natural objects, processing programs and acquisition of experiential knowledge (ibid.: 80). In my own comparison between healers and molecular biologists, I found these sensorial skills and forms of embodied knowledge to be highly reduced in molecular biologists and inhibited by attempts to minimize the feeling of materials in controlled manoeuvres and environments. The intimacy with 'indigenous medicine' seems to have nevertheless enhanced engagements of molecular biologists in the world in the South African context in particular, although investing more in transforming the RCT than envisioning innovation in indigenous medicine.

The very first thing the codirector of the South African branch of TICIPS explained to me was the need for a different kind of trial. The current 'closed pipeline' of the RCT needed to be filled with perforations that would maintain the research process 'in the inhabited world' throughout its four phases. He also explained that the other products that people consume in the everyday should be taken into consideration, since those products can inhibit the 'actions' of found molecules or medicines; he noted this with the example that the consumption of wild garlic inhibits the effects of a plant remedy against HIV/AIDS. A new model of clinical trial could override some of the current limitations of RCTs (Johnson 2011: 7), such as a translational validation model, as proposed by Patwardhan and Mashelkar (2009) for testing the efficacy of Ayurvedic remedies. Essentially, the model is a proposition for reversed pharmacology, for a pharmacology that reverses the routine 'laboratory to clinic' with 'clinic to laboratory' (Patwardhan and Mashelkar 2009: 806), and, in this way, a pharmacology that maximizes inspiration from 'indigenous medicine', or from 'life in context', even if limited to a clinical context.

> The translational validation model was more economical, more effective and more affordable. Phytomedicines had great molecular diversity and were very advantageous as therapies. They fell within the boundaries of drug-like properties, they had a greater number of chiral centres, an increased molecular rigidity and they follow the 'Lipinski rules of five',[5] which stated the smaller the better. (Johnson 2011: 1)

Recognizing indigenous medicine is thus clearly done through its plant molecules, aiming through this route to deal with human dignity.

It is within the new paradigm of molecular biology that indigenous knowledge is being revived and that a critique of the current RCT model is being made. This reveals a desire to bring more of the world into the laboratory. In fact, life in molecules, the plasticity of life the molecules behold, may blur the boundaries between the clinic and the laboratory. As the organism somewhat becomes the laboratory, the life that surrounds and permeates the organism becomes more relevant, as do the other beings it engages with, including plants. Knowledge may in this way break with a model that shuts out the world and include these other voices not as something to verify, test or regulate, but as worthy healing practices from which to learn. This would require embracing the entanglements healers delve into in healing and in rewriting life's story, as is somewhat done by molecular biologists, although without taking into consideration life's contingencies, such as the entanglements of bodies not only with foods eaten and a whole plant but also with skilled bodies, as well as with lived and felt histories and politics.

South African molecular biologists are highly aware of the politics involved in their research, not only in manipulating life at the molecular level, but also in manipulating life at the sociocultural, political and bioeconomical levels. Becoming a biochemist in South Africa is a way to acquire legitimacy in national and international meshworks. This is where the issue of African dignity is dealt with rather than within the science itself, even if it shows the potentials of doing so. Dealing with African dignity separately at the political level shows how both South African politics of indigeneity excludes indigenous knowledge from innovation and how the RCT is misadjusted to new developments in molecular biology. In either case, the RCT is described as deceiving because it is a process of closure to the inhabited world, and is hence unable to attend to the dignity of a people, or to the multiplicity of *A. afra* in life. The RCT's current model cannot account for molecular complexities in the environment, thus inhibiting its current most legitimized science in the process, namely, new developments in molecular biology.

Most scientists involved in TICIPS referred to the constant back and forth between laboratory and 'real life', and to how much is learned in the process. The desire to ground research in life opens towards the *isangoma* ways of knowing life, but then closes this opening as it turns these ways of knowing into molecular configurations with precise actions on a physiological process. What is maintained is a primary concern with 'life itself', which would have expanded its scope to broader ethics of living a good life but is restricted by a research model and agenda that is as narrow as it is powerful. While orienting 'life itself' is at the heart of the fundamental work of molecular biologists, it appears in this case to be limited to representing molecular processes in *A. afra* by its very desire to recognize indigenous medicine. In this move, rather than pursuing a project of rewriting life's story, as is done by Xhosa *izangoma,* molecular biologists refrain from doing so when under the banner of indigeneity, only to embrace it once again under the banner of innovation in science.

Improvisation

Medicines, or materia medica, are identified as such through various paths of legitimacy. The summit of scientific objectivity relies upon the RCT's empirical practice and is fully endorsed by biomedicine, strongly endorsed by much of humanitarian medicine, the FDA, the WHO and the NIH, and tentatively endorsed by traditional medicine. A still very dominating positivist view assumes that a well-designed RCT produces reliable evidence free of interpretation. The historical roots of the RCT's protocols and objectivity are masked through various implicit strategies, ones currently staging a biomolecular experiment as a solution to a tuberculosis pandemic, with a tangential goal of furthering humanitarian rights. The strong power/knowledge relations in which RCTs are crystallized make possible the lack of acknowledgement of the ontological underpinnings of this practice and potentially hinder fruitful negotiations with other ways of making medicines 'work' upstream. In its success, it is expected that following the procedures of the RCT will provide 'truth' about the medicine's efficacy regardless of bodies, places and contexts. When I began this research in 2006, I had initially titled my project in the Biomedicine in Africa Group at the Max Planck Institute 'Anonymous beholders of truth about the efficacy of medicine' because every time I enquired about how a medicine was 'known', I was referred to the RCT process. Although I later modified my title to 'South African Roots towards Global Knowledge' to bring some context to the 'anonymous' process, my research did confirm that it is in this view from nowhere that the RCT is still upheld, obscuring its embedding in ontological categories as well as its limited demonstrable specifics. Through this route, the RCT also holds a monopoly on what is named 'innovation'.

Scientific innovation in *muthi* appears to heavily rely upon distinguishing between primary and secondary qualities: 'primary qualities define the real stuff out of which nature is made, particles, strings, atoms, genes, depending on the discipline, while secondary qualities define the way that people subjectively represent this same universe' (Latour 2000: 118). Doing medicine can, however, be done through 'secondary qualities', namely, the visible, the lived and the felt, which have definite effects on the delineated primary qualities, as I have shown through Rastafarian *bossiedoktor* and Xhosa *isangoma* practices, as well as through my proposed phenomenological approach in anthropology. Continuously entangling and embracing or fine-tuning the senses can surely modify molecular configurations, including the ways these are perceived. Perhaps this cannot be done with certainty, but representing a mechanism and predicting a process in the laboratory will also be uncertain once reintroduced within life-making processes. As such, both routes, and all routes in between, work through improvisation, either in attempts to apply a distant model or in attempts to modify a certain state of being in-the-world.

While *izangoma* continuously improvise and 'read' the ongoing circumstances in terms of the healing initiation or event, molecular biologists improvise ways to enact a rigid model in practice, perhaps in need of even more refined improvisation skills, since they cannot go with the flow of ongoing life-making processes and are constrained to following a predesigned route. In the process of making things gel or cohere within the world, people 'are compelled to improvise, not because they are operating on the inside of an established body of convention, but because no system of codes, rules and norms can anticipate every possible circumstance' (Ingold and Hallam 2007: 2). The more distant the model to follow, the more need to improvise to make things 'work' within 'the specific conditions of a world that is never the same from one moment to the next' (ibid.).

RCTs are perhaps the most regulated type of research with the most specific guidelines in ethical policy statements. Taking this rigidity for granted when I began my research in 2006, there initially appeared to be little to discuss in terms of indigenous medicine, since the RCT design explicitly closes itself off to all assessments of efficacy other than 'the extent to which the physiological component adds significantly to the psychological component' (Lock and Nichter 2002). Working with indigenous medicine, however, leads to dissolving this division, knowing efficacy 'in life' or in the 'open air' rather than in controlled laboratories. Through time, I came to know RCTs as interstitial spaces where multiple ontologies come into being. While RCTs are often described as transnational networks, attending to how they operate in the world in their preclinical moment makes them look more like meshworks of people and things improvising to make things work in practice. Here, rules, codes and guidelines converge as well as diverge from life's contingencies, sometimes strongly enough to shake the design, such as in the case of proposals for new kinds of trials as done by the molecular biologists involved in the preclinical trial. The actors engaging in the preclinical trial are led to find relevance in the provided guidelines, but often these do not cohere with desires and logics on the ground, also exposing some of the moral consequences of claiming universal truth while dealing only with part of the issues at hand. The actors within the preclinical process I followed nevertheless mend the world together as best they can, adding reality to or improvising the preclinical procedure by, for instance, paying attention to how the medicine is prepared by the healers and how the plant is preferred in the everyday, all elements feeding into the experimental process, even if they are screened out in its results.

The preclinical trial I followed was filled with ontological improvisation. At first I thought I was straying off topic when assisting a drum session or a healer's initiation, but it became clear that this was how knowledge with plants was being made, namely, in the midst of life-making processes, prioritizing the lived and the felt. These elements in life are, of course, omnipresent throughout the clinical process as enacted by the scientist, but with the intention of minimizing them; the plant and the disease must be controlled and externalized, and the doctor, nurse and patient must be blinded or tricked into not knowing whether placebo or medicine will be administered. The vacuum-sealed laboratories, the gloves and garments, the machinery, the computers and, most of all, the strict model are meant to prevent leaving traces of the researcher's presence. In these ways, the medicine being done undergoes extreme transformation in the hands of laboratory scientists. The main paradox in this design is found in the claim that these procedures do not affect the nature of the plant and only reflect it, while still claiming exclusivity in 'innovation'.

It is through these very procedures of withdrawal that the RCT creates new ways of being in-the-world as well as new entities of beings produced biotechnologically in controlled environments. All the controlled measures create an apparent inability to grasp indigenous medicine, which defies causal logic. Perhaps the most mutilating aspect of preparing an indigenous medicine for an RCT is when the healer's presence is excluded, leaving no trace. Disregarding the 'placebo effect' and blinding both taker and giver of the medicine, disregarding essentially any relational effect in the ways of administering the medicine, are explicit acts of dismissing the world, the senses, the 'sum' of Descartes's cogito. While this also takes away the intuition mentioned by molecular biologists, it does not take away their legitimacy to exclude these aspects of research; on the contrary, it provides us with solely disjointed and often disconnected information in which meaning is difficult to make for the nonexpert in the field. Further, the RCT as currently designed does not inform us of the efficacy of the medicine in the world but improvises new things to be brought into the world, which in turn modify and produce the world. What we observe is not nature itself, but nature exposed to our method of questioning (Heisenberg 1971; see also Capra 1999: 40). The RCT is a highly deductive research process, a model and theory in search of data. Its results tell more about the model and the humans behind it then about how a medicine is useful in the world. What it tells about the humans behind the model is that they do not want to be there.

The ontological divide between nature and culture as currently set within RCT protocols, and as enacted by the actors, often to their dismay, is what excludes the healers and their improvisational skills as forms of innovation under the indigenous banner. This sort of exclusion is not new, nor restrained to 'indigenous healers'. In the 1950s statisticians and molecular biologists were not welcome in the RCT process, but they have now become some of its most indispensable actors. When this occurred, the RCT also changed. The doctor's experiential knowledge was set aside and patients lost most of their agency (Marks 1997: 187); the superiority of "objective" measures displaced themselves from evaluations anchored in the incorporated abilities of the practitioner towards the clinical status of patients (such as laboratory tests results) (Marks 1987, 1997, Löwy 2000). It is this very struggle that reemerges in the encounter between molecular biologists and Xhosa healers. Healers are currently accessories in the preclinical process, providing hints at what molecular configuration to look for in the plant according to how indigenous actors prepare the plant and, perhaps most importantly, ensuring compliance and the participation of local people when the clinical phases begin.

It is, however, never imagined that taking the burden, the toil and trouble of life, upon oneself is as much a part of creating possibilities for acting upon life as is access to the latest life-enhancing biotechnologies. It is, perhaps, not only the latest high-tech biotechnologies that enable infinite possibilities to act upon life in meaningful, real and creative ways, yet those who improvise by doing without them and entering into more intricate relations within mediums in the inhabited world, might have this capacity as well in a much broader sense.

Most scientific experts state a desire to learn from the healer's practices, yet they cannot find a place for indigenous actors to stand, as they are not recognizable experts, nor part of the RCT grant design and proposal. The process of one RCT can last up to seventeen years, leaving plenty of time to disconnect from the world outside the clinic and laboratory. This time and investment can justify claiming innovation as it loses its links with indigenous knowledge through the process, while 'indigenous knowledge' cannot claim innovation by doing something new with scientific 'discoveries', a process usually shunned or even made illegal. Life nevertheless continues its course, and the ontologies brought into being acquire new meanings both inside and outside the experiment, showing the process of making medicine to be multiple and open-ended. A project moving away from discovering the 'laws of nature' and moving towards ways of composing life points to a need to enhance improvisational skills in movement.

The context of postapartheid South Africa and the consequent revival of indigenous medicine somewhat permeates the work of molecular biologists implicated in the preclinical trial I followed, a common orientation that seems to further converge with the recent turn within the science of molecular biology as forms of improvisation. I have pointed to potential connections between life found through feeling sounds in *isangoma* practices and life found in molecules in molecular biologists' practices, as they both imply dealing with life as a movement of opening. As such I suggest that finding worthy ways of living with the real needs to be improvised by problematizing 'life itself' rather than reifying it. This proposed direction of weaving molecules into life upstream and into the flows of life rather than solely downstream can also be strengthened by learning from Rastafarian *bossiedoktors*. With an explicit aim to disband colonial rule, the Rastafarian movement was founded in its very opposition, taking up most of its secondary qualities and maximizing legitimacies in the visible, the lived and the felt in a quest for equality for all mankind. Rastafarian 'One Love' thus appears to take up precisely what the current One Health Initiative fails to address. The recent One World–One Medicine–One Health initiative unites animal, food and human studies to join forces through findings of commonalities between species, doing so at the molecular level. The Rastafarian 'One Love' movement also aspires to universality, but in the context of the lived experience of being in-the-world rather than of a shared 'one nature'. Rastafarian 'One Love' might thus profile the shadow side of the One Health Initiative. Somehow these two movements seem to fulfil the other's blind spot, showing another potential area of convergence towards imagining a common world. The conclusion aims to tease out more hopes and kinds of life that partake in the preclinical process, even if sometimes solely in its shadows. As such, how the RCT is imagined is basically what defines the limits of improvisation in doing medicine.

Notes

1. For some scholars, it was Darwin's evolutionary view of life as a distinct object of study in its own right that gave rise to the modern notion of life itself. For others, it was Aristotle or Descartes who divided the two domains of life, mind and body. For others still, it was Claude Bernard who paved the way for the 'experiment' and its subsequent division of nature and culture. In the midst or aftermath of Arendt's (re)introduction of the notion of 'life itself' in *The Human Condition* (1958), similar other notions have emerged. In *Homo Sacer*, Agamben (1997) has, for instance, introduced a notion of 'bare life' as instantiating the fatalistic direction towards which biopolitics ultimately lead. The refugee camp is the eventual result of these kinds of politics because of their exclusive concern with vital necessities (life itself), realizing what he names the most absolute *conditio inhumana* (Agamben 1997: 179), as this condition omits consideration of broader aspects of life. For Rose, the hierarchy between *zoë* and *bios* would be dissipated in the current molecularization of life: 'Our very understanding of who we are, of the life-forms we are and the forms of life we inhabit, have folded *bios* back onto *zoë*. By this I mean that the question of the good life – *bios* – has become intrinsically a matter of the vital processes of our animal life – *zoë*' (2007: 83).
2. The biotechnological enterprise itself with its idea of scientific control of life can however be traced back to events that took place well before the

discovery of the double helix of DNA by Watson and Crick in 1953 (Pauly 1987). In a general sense, domestication of plants and animals as well as agricultural technology are efforts to transform living nature for human purposes. From its beginnings one aim of medicine has been a particular form of control of life, namely, the restoration of a state of health when normal function is disturbed. The Scientific Revolution reinforced these impulses; 'the alteration of plants and animals was part of Francis Bacon's plan in the *The New Atlantis...*' (ibid.: 4). By the middle part of the nineteenth century as the importance of physiology to medicine grew, manipulation of organisms greatly increased within the medically defined polarity of health and disease.

3. This great biopolitical technology – anatomic and biological, individualizing and specifying, directed towards increasing the performances of the body, with attention to the processes of life – characterized a power whose highest function is to invest life through and through (Foucault 1976).

4. TICIPS used various cutting-edge scientific methods to understand *A. afra* and *S. frutescens*, such as high-pressure liquid chromatography, 1-D and 2-D nuclear magnetic resonance, liquid chromatography, mass spectrometry, X-ray crystallography, firefly luciferase bioluminescence and fluorescence microscopy (Johnson 2011).

5. The Lipinski rule of five is a rule of thumb to predict the absorption of compounds in humans as well as the drug-like qualities of a chemical compound. The rules apply only to absorption by passive diffusion of compounds through cell membranes and point towards a certain weight and number of groups in the molecule. It is also, interestingly, known as the Pfizer rule of five, showing its clear link with the pharmaceutical industry. Pfizer is one of the largest research-based pharmaceutical companies, founded in 1849 by two recent German immigrants to the United States, Charles Pfizer and Charles Erhart.

Conclusion
Imagining the Clinical Trial

Knowing *A. afra* is done in mediums through varying roots and routes, in controlled environments and in the 'open air', learning about and from the plant, through ancestors, dreams and sounds as well as in attempts to transcend contexts to heal universal biological bodies. The clinical trial process can be imagined as tending solely to 'life itself'; however, this is not done in practice, nor is efficacy of a molecule in a human cell ever fully disentangled from broader life-making processes in-the-world. I have shown how numerous problems arise during a process aiming to reconcile the search for a worldly molecule with retrieving the dignity of a people. In this conclusion, I aim to focus on the problems that arise in immediacy and that touch as much on indigenous medicine as on the search for a medicine that initiated the preclinical trial.

Converging as well as competing hopes are part of what keeps the meshworks of making medicine alive (or not). Actors connect around a shared concern with human life, even if they approach this concern in different ways. I have evoked numerous lines of hope that permeate the preclinical trial: hope of 'discovery'; hope for a cure; hope to enhance life's performances; hope to save lives; hope in scientific innovation; hope in 'nature'; hope in humankind; hope to show to the rest of the world how indigenous medicine works; hope in 'wild' plants; hope for the dignity of a people; hope for a better life for some; hope to heighten a career for others; hope for recognition; hope in healing a nation. On many different levels, I found hope. Hope is here not understood as something in the mind, a driving force or a film of psychological cement binding the biophysical and sociocultural together. Nor is hope an abstract cultural force that will explain what is going on. It does not have an intrinsic capacity to tie people and things together in a project. Leibing and Tournay

(2010: 5) similarly argue that 'technologies of hope' are not decreed a priori, but rather emerge through everyday entanglements between people and things, often then becoming tools of governance, or at least pointing towards lines of becoming. I have found that engaging in the process of the preclinical trial is filled with motions of hope that emerge at personal, local, national and transnational levels.

In the preclinical trial of *A. afra* I followed, the mingling of ontologies makes the process of generating biopharmaceuticals complex and fragile, but still possible. While imagining the RCT requirements is a source of dissonance with indigenous practices, it also offers a map or a guide to follow that provides occasions to learn *from* indigenous medicine, rather than solely *about* indigenous medicine (which remains primary). Imagining African roots or indigeneity also pulls away from simply doing medicine in line with what is going on, offering guidance rather than a guide. I thus invite you to imagine what it would look like to really take this guidance into consideration and reach some forms of indigenous dignities in the process, namely, by exploring what would happen should felt sounds in life be considered as hierarchically primary qualities rather the current molecules 'out there' or even better, considering them as equally at work. Doing medicine means partaking in life-making processes as an inhabitant in-the-world before any attempt to follow a map or to accept guidance can be made. The question of the efficacy of *A. afra* is found to be woven into lines or overlapping motions of hope, in particular those that align themselves along the thrust of humanitarian efforts to 'save lives', of the 'African Renaissance' and of 'indigeneity'. This assessment leads to broader understandings of how experience is continuously organized in-the-world. In this way I begin to tease out my own hopes for future directions in anthropology as they emerged throughout this research and journey.

Saving Lives

'Saving lives', as evoked in humanitarian efforts, is to save life itself from death in physiological and physiochemical terms. A large number of organizations, both legal and ethical, are all working in concert to find medicine for identified diseases of all sorts, situated in a naturalist ontology of rational man as external to the world, which can be discovered as well as, more recently, innovated upon. In this line of becoming, humanitarian aid aims to 'save lives' with life-saving technologies, and the RCT is the corresponding medium to achieve this. While therapeutics may have initially led clinical studies and made ways for better lives, the centre of attention seems to have shifted to the demonstration of a method, more specifically, a demonstration of precise physiochemical processes at the molecular level. In opening potentials to orient life, the quest for new molecules opens some interests in indigenous medicine or in 'naturally' occurring phenomena, although mostly for inspiration in creating new entities in the laboratory.

In this short dip into healing with plants in-the-world, concerns with efficacy also appear to have moved towards an emphasis on 'safety': worries about safety ('It is safety that I'm worried about'; 'We are just making sure it is safe') were repeatedly stated by the American director of TICIPS who clearly held a strong belief in the necessity of RCTs to regulate plants on the market for the good of the people. He further supported the role of the NCCAM as a regulatory agency trying to protect the public, with the legal ability to control the use of plants in the everyday. In concert, collaborating organizations aim to 'conserve nature', or protect it, often from humans who destroy it, at the same time as it is made possible to do so by ontologically separating humans from nature. This contradiction is weaved into a hope in finding truth and innovation to save and protect lives through following the steps of well-designed RCT models.

This order of legitimation is delineated, followed and upheld by the scientists involved in the preclinical trial I followed, some of whom play on all three fronts (molecular, local and transnational), yet it is an order of legitimation mostly upheld by boards and distant agencies, such as the WHO, the FDA, the NIH, ethical committees and, more recently, the clinical trial monitors, which enforce a sole preoccupation with 'natural' life as separate from culture. In this hope to save lives, vital biological mechanisms are primary, while qualified life, life in context, is secondary, if not altogether absent. The humanitarian logic that permeates this order of legitimacy is that if one remains biologically alive, then other forms of life can thrive. How to intervene is currently understood at the molecular level, through medicine, which becomes the route and commodity to 'save lives'.

Lives are to be saved with biopharmaceuticals. James Orbinsky formulated this very rationale in his Nobel Lecture as the president of Médecins Sans Frontières at the time:

> Some of the reasons that people die from diseases like AIDS, TB, Sleeping Sickness and other tropical diseases is that life saving essential medicines are either too expensive, are not available because they are not seen as financially viable, or because there is virtually no new research and development for priority tropical diseases. This market failure is our next challenge. The challenge however, is not ours alone. It is also for governments, International Government Institutions, the Pharmaceutical Industry and other NGOs to confront this injustice. What we as a civil society movement demand is change, not charity. (1999)

Change is demanded in making the drugs available, not with regards to the ways these are to be made, nor with a critique of the particular narrowness in the kinds of lives to be saved. In the humanitarian endeavour, the kinds of lives to save, for whom, when and with what kinds of realities are left untold.

Lives are saved as a right, a right to health, which is, essentially, a right to be alive. Instead of saving souls, as was done in previous humanitarian missions, saving biological life has emerged as an incontestable good upon which human rights are also fundamentally anchored. The 1948 Universal Declaration of Human Rights roots its legitimacy in the belonging of all persons to the same human nature. It remains difficult to criticize the good of this right. However, in following where efforts are placed to tend to this right, we come to understand that it relies upon a particular privileging of biological life above numerous other kinds of life. This completely dismisses appeals to indigeneity, which clearly fall outside of its immediate concerns, notwithstanding its disregard for the active body in the world in both healing and in making life liveable. In this right to live, medicine is reified as a life-saving device containing health. Access to it is 'essential' and death is often attributed to a failure of the market to have provided this access (Laplante 2003). This, I argue, begins in the ways of making medicine, where only certain forms of life are known or want to be known, as shown through this study. As such, investments are made in particular forms of life and not others.

Medicine, in this approach, is a closed object with well-defined mechanisms and that can travel and heal. Molecular biologists are aware of the multitudes of possibilities of action that the same molecule can entice within bodies (both human and nonhuman) in the medium. It is more or less acknowledged that even a precisely defined dose of a plant prepared in laboratory studies and passed through a successful RCT cannot have predictable effects when it goes back into the world; bodies will only be healed should they respond accordingly through the envisioned mechanisms and effects. It is also known that indigenous ways of healing can enhance the chances of making a plant 'work'. The 'humanitarian' objectives of the RCT are, however, set on solving the tuberculosis pandemic in molecular terms, which takes precedence over indigenous ways and dignity, finding the pandemic more important to deal with than the dignity of a people. Ensuring the continuance of vital functions is thus primary to ensuring people have a life worth living. Tuberculosis is attended to as a 'neglected disease' while a neglected people suffering from the disease is not attended to per se. Hence, the dream to redress African dignity through an RCT may not turn out as hoped. It may instead become 'randomized controlled crime' (Adams 2002), or the criminalization of traditional practices, should RCT results 'disprove' the efficacy of a plant through its externalized lens. In the end, the molecule carries the burden of truth about the efficacy and safety of a medicine, and this is closely monitored by distant RCT protocols. Molecules become both moral and political.

The RCT is hoped to find a solution, a molecule or precise plant preparation in this case, that will mitigate a humanitarian problem; however, the route followed is to save biological lives through biopharmaceutical innovation, both stripped of their humanities. The preclinical trial of *A. afra* studied did not dissociate itself from this approach in its design. In practice, however, numerous researchers in TICIPS bend the line of the clinical process for it to fit the context. South African researchers, for the most part, know healers and their legitimacies very well. Quite often their mother, uncle, grandfather, daughter was a healer. Some explained how they heavily relied upon knowledge emerging with *muthi* to best target the most promising molecular configurations. However, it seemed perilous for these researchers to establish these kinds of bridges in their science; such bridges were not officially required, and even were asked to disappear from the research. The researchers' discourse was generally more imbued with their scientific expertise, stating, for instance,

'I am a consultant, I do the problem solving, I am a pure research expert' or, 'It's fundamentalist work.' In this way they did not doubt the perceived benefits of making a biopharmaceutical and only shyly engaged with indigenous medicine in their science. In addition, politics currently reign in postapartheid South Africa, and there was no possible way to extract myself from these issues, no more so than the process of the preclinical trial managed to remain 'objective' and a neutral, incontestable good in this context. Making medicine with *muthi* is complex territory. Histories are very much alive, and the imposition of a single standard of legitimacy is not a subtle form of the colonization of bioresources and the people inhabiting them.

Much more could be learned from the interlacing of natures and cultures; however, these ways of knowing and being mostly escaped the preclinical trial even if they partly slipped into hopes of reengaging with African roots. This explains some of the ambiguities found at the national South African level, as well as at the transnational level. TICIPS announced the recognition of *muthi,* but without learning how to do so. Revived interest in *muthi* may, however, foreshadow a reconfiguration of the RCT to yield place to multiplicities of cultures and perhaps also of natures. If, as Merleau-Ponty (1964) states, each act moves towards the indeterminate, or towards a world that we have not yet configured and that will only make sense the moment we engage with it, than this engagement, this ability to enter in relation with the shared world with plants, can become as important and useful as the standardized clinical efficacy of a plant tested in the laboratory.

The clinical trial model offshored into a new context such as South Africa signals a complex, plural world and, at the same time, brings into question some biopolitical foundations as well as their objective premises. If, as Küchler (2006) suggests, technological innovation is oriented towards smart materials transcending the mind-body, nature-culture dichotomies, transforming meaning into matter, then we would have to change the paradigm to follow the rhythm of innovation, namely, in opening the current RCT, which holds a promise of closure. This implies thinking at the interface between mind and matter. It appears like new molecular biology is moving in this direction, a space explored in life by *izangoma,* who excel in ways to transform meaning into matter. Stroeken (2008) finds in the synaesthesia of the senses, used by healers, a radical transformation in the ways in which bodies and senses enter into relation with the environment. This gives life to things, inhabiting things and understanding them in mediums as much as in controlled laboratories. This is not to say the laboratory is not necessary, but that it is merely a voice within others. I have shown how it cannot hear the healing efficacies of sound, nor can it deal with the plant's vitality, let alone be preoccupied with the vital necessity of needing to have a life worth living, as clearly enacted through Rastafarian *livity*. Powerful hopes in African dignity move through the preclinical process, pulling its narrow notion of life itself onto broader horizons around issues of indigeneity; however, they do not seem fully able to fulfil their goal through the preparatory process, since the RCT is ultimately to be done in a humanitarian hope in 'one nature'.

African Roots

Nelson Mandela's recent passing echoed throughout the world, showing a man and a politician who will be missed for his uplifting sense of humanity. His powerful gestures have moved the world and given hope to an entire people, transforming their everyday lives, and thus have played an important role in healing the nation and beyond. In the South African context the recognition of *muthi* is linked to this hope for human dignity. In his inaugural speech as the first black president on 10 May 1994, Mandela appealed to experience as a way to move forwards: 'Out of the experience of an extraordinary human disaster that lasted too long, must be born a society of which all humanity will be proud' (Mandela 1994). It is also through the experience of daily lives that he invited South Africans to produce a reality that would reinforce humanity's belief in justice, namely, through engagements in the medium, stating that '[e]ach time one of us touches the soil of this land, we feel a sense of personal renewal' (ibid.). Soil becomes the grounding for human dignity, one shared with trees and *muthi*; Mandela appealed to this metaphor to convince a disillusioned people that they indeed belong: 'I have no hesitation in saying that each one of us is as intimately attached to the soil of this beautiful country as are the famous jacaranda trees of Pretoria and the mimosa trees of the bushveld.' (ibid.) More than Mandela's words, it is his voice that has been powerful in immersing others in his meaningful and transformative sounds.

Mandela's motions to move people towards a better life were a few years later formulated as an 'African Renaissance', a workable dream, one that, however, became less grounded in immediacy and more in envisioned futures. This move was translated similarly within the preclinical trial, which evoked indigeneity and African roots as a distant horizon, as what has become the humanitarian horizon, although with the important distinction that it is less about 'one nature' and more about a distinct nature-culture dichotomy, either an indigenous one or a scientific one, made of comparable yet separate entities of existence that could collaborate. The ontological premises of a world divided between natural and cultural components are fully in place, making them slightly more amenable to scientific experimentation. The preclinical trial of *A. afra* was brought into being in part through an idea of such a thing as 'indigenous medicine' in South Africa, one nevertheless inseparable from a particular nature. In other words, in the recognition of an indigenous nature known by an indigenous culture, the ontological premise of naturalism is transposed within its own ontological divide. It places the indigenous/nature 'object' as 'culture' in the naturalist paradigm, maintaining the dominance of primary qualities in the sciences, explaining in part its exclusivity in innovation. In this way, indigenous medicine is known from the outside, never really entangling with it as real ways of knowing and healing.

This is the way that the preclinical trial emerged as an NCCAM project, as a complementary and alternative practice to be verified scientifically. In the South African context, indigenous medicine is, however, not complementary and alternative, but thoroughly anchored in everyday lives as well as part of the dignity of a people. As such, the multiplicity of indigenous medicine is regrouped under one *muthi* that can be nationally reified. The preclinical trial of the indigenous plant *A. afra* played right into these hopes concerning recognition of a knowledge, recognition of a medicine and recognition of a people, which are seen as embedded in a single nature linked with a single culture, in this case, the culture of a unified South Africa and its indigenous nature, also multiple or 'biodiverse'.

The actors directly implicated in the preclinical trial hoped to see *A. afra* travel in the global markets while keeping its national South African emblem. Some *izangoma* initially joined this process, sharing a hope of contributing to the common good of the nation and beyond. An ambiguity, however, persisted. A project entirely financed and standardized by the NIH in Washington, D.C., translates, in South African histories, into a colonial act. This has been contested in the national initiatives running clinical trials. The actors of the MRC, and more specifically of its IKS branch, object to external financing in promoting the motto of 'African solutions for African problems'. Matsabisa, the director of the IKS branch of the MRC, amongst others, explicitly refused to conduct Phases III and IV of the RCT model. In this way, the medicine is unable to travel world-wide, and national benefits are guaranteed rather than overseas benefits (Matsabisa 2008). Matsabisa criticized TICIPS for accepting external funding. Perhaps sharing this critique, after obtaining $4.4 million on 18 July 2005 in the NIH World Cup of Science competition, TICIPS pursues its mission today through new sources of financing. In 2011, TICIPS created a consortium of universities, the Multidisciplinary University Traditional Health Initiative (MUTHI), which obtained €2 million in the World Cup of Science in Brussels. In order to get out of this dependence on external financing, public-private partnerships are envisioned, as well as a proposition of a new model of clinical trial. The new model of clinical trial proposed by TICIPS is a reversed pharmacology model, carrying hopes that more of indigenous medicine could enter the clinic and filter through what it will be narrowed down to in laboratory studies. At the same time, the model consists of new forms of medicine at the cutting edge of biopharmaceutical, bioinformational, biomolecular, biotechnological and biomedical innovation, thus hoping to make the best of both worlds.

In this process of translating indigenous medicine into biomedical language and practice, what occurs is a disappearance of the 'indigenous'. TICIPS was renamed The International Center for Innovation Partnership in Science (TICIPS) in 2010. It seems that the 'indigenous phytotherapies' appellation in TICIPS (originally The International Center for Indigenous Phytotherapy Studies) subsumed under the banner of 'innovation in science'. This move is described as one from translation of IKS to innovation for the bioeconomy, thus liberating indigenous medicine from stagnation in the past as well as moving 'beyond the custodians of the knowledge' (Johnson 2011: 1). Perhaps this will also avoid the problem of otherwise recognizing indigenous medicine in a politics of exclusion, or perhaps indigenous medicine will simply become 'science', as has been the case with other RCTs. In both cases, hope in one culture/one nature appears to hold a compromise. It is how the preclinical trial of *A. afra* came into being in the first place, whether or not the result will be successful in recognizing indigenous medicine if *A. afra* travels in global networks. These hopes are at the interstices between the humanitarian and indigenous horizons, as attempts are made to equate a plant from the first domain (nature) with its belonging in a second (culture). At least initially, an indigenous medicinal plant is positioned to transcend at the same time its own biodiversity in its relations with humans and mediums, as well as its South African indigenous diversity.

Historical events of health and healing in South Africa are interlinked with colonial and postcolonial histories. The legal processes instituted against *muthi* went deep into the political nature of *isangoma* practices: '*Muthi* was in itself deemed responsible for political success or failure. Acquisition of proper *muthi* and the powerful doctors who administered it were considered essential to a chief or king's rise to and maintenance of power' (Flint 2008: 71). This makes it clear how the imposition of an external legal standard to test *muthi* is a form of colonization of local power/knowledge, even nowadays. Even if in the preclinical trial of *A. afra* there could have been questions of the usurping of intellectual property rights linked with indigenous knowledge concerning *A. afra,* this element did not emerge as a problem. This might have to do with the common 'nature' of *A. afra* across regions and traditions, not linked with a single group. Hence, *A. afra* sneaks in more easily within the politics of natural resources. *A. afra* is thus claimed at the national level, appealing to the sui generis protection of traditional knowledge (Collot 2007). This recognition at the national level is another reification of the plant's primary biological life at the expense of the many other mediums in which it is alive.

While issues of intellectual property rights did not emerge as clearly problematic for the healers, the issue of patents did come up a few times during my research. The issue of patents arose around the recent case of *Hoodia gordonii* cactus 'discovered' by the San (bushmen of the Kalahari) and patented without notification by Pfizer in 1986. This story ran throughout the country and was only recently resolved, at least in part, following long legal processes (G. Thompson 2003). The South African Council for Scientific and Industrial Research (CSIR) is the research institution that patented the drug, granting United Kingdom-based Phytopharm a license and also collaborating with the pharmaceutical company Pfizer, who released the rights in 2002. While the 1992 International Convention on Biological Diversity (CBD) recognized the contribution of First Peoples groups to drug 'discovery', and the 2004 South African Biodiversity Act provided a blueprint for how researchers were to compensate local communities in the country, the history of hoodia as a 'success story' for benefit sharing with the San Bushmen remains ambiguous (Osseo-Asare 2014). Further, the stories told still let on that the use of *H. gordonii* is restricted following its patent, something that a Cape Nature agent confirmed, to the astonishment of the *izangoma* and myself. The patent on *H. gordonii* first of all only concerns the use of the plant as an appetite suppressor, while the indigenous uses of the plant are much wider and varied. Second, the patent only concerns a specific process of extraction of a very specific bioactive compound of the cactus (substance P57) for a specific use, which in no way detracts from the use of the cactus whole as found in the 'wild'. However, the resolution of the case states that its sale in the form of dried material can constitute a violation of the patent, hence making its sale illegal. Some lawyers seem to interpret only the intention to sell *H. gordonii* in the market as sufficient evidence of the violation of the patent (Gruenwald 2005: 29).

The question of permits to collect plants solely allocated to associations of people and not to individuals was also raised as an issue, as well as the question of control of territories to create parks and nature reserves. These two issues greatly affect the *izangoma* and the *izinyanga,* as the inaccessibility of plants constitutes a threat to their survival. In asking an *isangoma* which solution he could foresee with regard to limited access to *muthi* and, consequently, the lack of opportunity in knowing them, he answered that he only needed a small parcel of forest, any parcel, and within three years, he could recreate 'live knowledge' in engaging with plants, thus providing a way to sustain the kind of knowledge that the reversed pharmacology model wishes to benefit from. Navigating at the interstices fluctuating between legitimation of ancestral practices, where nature is part of people, and legitimation of the process of standardization of nature into a biocommodity is a dizzying process.

The commerce of South African medicinal plants is an industry in full expansion, representing 5.6 per cent of the national budget in health (Mander et al. 2012). Drugs derived from plants such as paclitaxel, vincristine, vinblastine, artemisinin and camptothecine have generated $65 billion in sales up to 2011 (Johnson 2011). With its 27 million consumers, the commerce of 'wild' plants is vibrant. 'Natural' products have contributed to around half of the small molecules approved during the last decades (Patwardhan and Mashelkar 2009: 804). The age of 'blockbuster' drugs is perhaps diminishing, yet the frenzy around 'wild medicinal plants' is reviving and thus needs to be monitored. The ethnobotanist from TICIPS mentioned in chapter 3, expressed a corresponding need to look at combinations of compounds as used in the everyday, which often make them more efficient. A new legal authority could then ensure their safety and security, one she referred to as 'medicinal plant security'.

The emergence of an international standard for the sustainable collection of medicinal and aromatic 'wild' plants such as the Medicinal Plant Specialist Group in 2007 testifies at once to the expansion of this market as well as to the will to master it within the same terms as the RCT. The Medicinal Plant Specialist Group, in fact, uses the RCT as a tool of legitimation for the safety and efficacy of 'natural' products for human health. The RCT is maintained as the legislative device and research method to ensure the legitimation of the safety and efficacy of 'natural' products from which to innovate. Such organizations as the Species Survival Commission, part of the International Union for Conservation of Nature, based in Switzerland, ensure the application of the international standard. The Species Survival Commission, like the RCT, centres its efforts on biological life, placing qualified life second, and not interlacing the two. The emergence of such a standard minimizes an understanding of the ways through which biological life and qualified life interweave and meet poetically, safely and efficaciously.

In the dreams evoked to push the preclinical trial forwards, hope dwelled in molecules that would maintain their indigenous roots, a dream also initially shared by a few *izangoma* connected to the trial. In this organizational performance, A. *afra* was nevertheless reified in the transnational concern of finding a cure for tuberculosis, with the hope that Africa, South Africa or indigenous knowledge would be recognized for this innovation. These multiple hopes were crystallized in the preclinical trial aiming to demonstrate A. *afra*'s efficacy in a way that would be recognizable within an RCT, perceived as an access point or an opening to the 'rest of the world' through indigeneity. It is, however, a 'humanitarian' motion of hope that ultimately defines this legitimacy, one that remains powerfully challenged through everyday practices that in the end make this 'work' (or not) in life-making processes.

Healing Roots

Tylor speaks of fieldwork in anthropology as 'an aesthetic integration that will have a therapeutic effect' (1986: 125). This can be done by provoking a transformation through deepened engagements in the environment, for instance, in engaging with a plant as a means for healing. In other words, it is hope of finding the benefits in plants through proximity, feeling them, like kin. 'It is, in a word, poetry – not in its textual form, but in its return to the original context and function of poetry, which, by means of its performative break with everyday speech, evoked memories of the ethos of the community and thereby provoked hearers to act ethically (Jaeger 1945: 3–76)' (ibid.: 125–26). Hope in poetics moves through the RCT in a number of ways. It appears more explicitly in both *izangoma*'s and *izinyanga*' ways of engaging with medicine, which rely upon profound aesthetic integration in-the-world. It appears as well in one of the pharmacologists' laboratory, who has made this place his home as well as his hope of conducting good research by involving himself thoroughly with the materials. It may also appeal to molecular biologists following the path of science as a way to surpass lived inequalities during apartheid, somewhat following in the footsteps of Mandela's powerful voice, gestures and performances that moved towards reestablishing the unity and equality of a people and place. It appears in the sophisticated manners in which healers delve explicitly into the preobjective in order to enhance life or to heal. Rastafarian *livity* seems to be all about maintaining this state of simply being alive in a 'good way', and it turns out to be a daily struggle. Xhosa *izangoma* have developed intricate events and performances to reach deeper into this state, making it an art and a science. In these practices hope dwells in the land, one that is not 'used up', one with which good relations need to be continuously weaved.

Initiation into the divinatory art is first and foremost an embodiment of 'sounds', learning how to intermingle with the world. 'Natures' are musical, interactive and aesthetic, indissoluble from 'cultures'. To make a medicine useful, it requires enhancing one's sensibilities. Developing sensitivities by ridding oneself of standards and by boosting the organic link with the world legitimates knowledge and can lead to healing potentialities. In this context, *umhlonyane* can be placed under the bedsheets of the person 'doing *ngoma*' to 'purify' that person for the next day's healing/learning session. It can be taken for the stomach, cough, fever, to clarify dreams or to move spirits around. What is important in the collection of *umhlonyane* is not its calculated biological dose or composition as much as its qualified organic life: how, where, when, by whom and in which ways it is collected in relation with the (dis)ease it is meant to alleviate or reorient. 'Life' through the plant remains intact or useful when particular relations between people and the plant in the shared medium are fine-tuned. The particular engagement in-the-world lets the *isangoma* activate the plant in its vitality in-the-world. This world or medium is that of histories, politics, ancestors, earth, gods and the lived experience in proximity with the plant and those engaged in the performance. Those enactments play themselves out through aesthetic or poetic entanglements of natures/cultures, opening to multiple possibilities of orienting life through life.

Izangoma and *izinyanga,* on the one hand, enact their relations with medicines when asked how a plant works and when answering through gestures and motions, as well as when expressing their preoccupation with fruition or *livity* in the everyday. While *izangoma* do not bring into being a divide between 'nature' and 'culture', Rastafarians have a notion of 'Ital' or 'I-tal', meaning organic, natural, pure, essentially, all that is provided by Jah which contrasts with what is considered as transformed artificially through technological manipulations rendering it 'lifeless', as also discussed in chapter 5. The notion is applied to foods, with a prohibition on salt, pork, some other meats and on processed foods generally (Yawney 1978: 203). 'I-tal' also refers to smoking *dagga* pure (ibid.: 183) as opposed to with tobacco. Healing plants are those found through good relations, with care and the proper skills, some being more powerful if closer to Jah. In this sense, like food, plants can be both 'natural' and 'cultural', entangled. There are no primary qualities superseding secondary qualities, since all are to be felt, only some more intensively than others. Further, as with the Xhosa *izangoma*, Rastafarian *bossiedoktors* are not preoccupied with the objects perceived to heal in themselves, nor are they preoccupied with the problems of health and healing in themselves, 'since in their approach to life, health will follow as a corollary to a certain kind of existence' (Yawney 1978: 199); a life worthy of living thus ensures 'health' rather than the other way around, as found in biomedicine.

Further, it is not the plant that is 'natural' (or not), it is the engagement with the plant that can be more or less aesthetically integrative and therapeutic. A plant in this political economy is not a bioresource, nor is the environment a 'reservoir' of the latter; the environment remains inhabited by humans. Hope in poetics might be shared by biologists and sentient ecologists who feel part of the environment. It may be understood in this ontology that 'humans damage places not because they fail to understand them, but because they are yet to feel for them, like kin' (Wattchow 2012: abstract). Proximity with the environment, skills or this way of 'being with' plants in life is a way to control potential hazards, like travelling standards and protocols, albeit in different ways currently unrecognizable within the RCT. It is in the relation of humans in the medium that indigenous knowledge with the plant emerges, since the condition of being human is fully acknowledged in knowing. In this regard, the clinical trial remains limited for both healers and scientists. Most scientists involved in the trial, in fact, mention this limitation with regret, yet other motions predominate.

While the RCT is a gold standard to make biopharmaceuticals, it seems that *muthi* may follow the standard to resist standardization, creating dissonance. The respective ways of doing medicine almost reverse the order of 'things': the RCT proceeds with a model of patterned regularities or standards through which it filters experience in the world as excess to dismiss; 'doing *ngoma*' and Rastafarian *livity* deepen experiences in the world to dismiss current regular patterns or standards of the everyday in order to make way for new orders. The standard to avoid standardization can also be read as forms of resistance to colonial rule, which is explicit in Rastafarian practices and philosophy as well as omnipresent in *isangoma* practices surviving in the shadows. Similarly, the RCT may be read as a form of colonization through the standardization of indigenous knowledge or of all nonexpert scientific knowledge, as well as of 'bodies', 'nature' and 'life', which are made into objects of enquiry. However, *isangoma* healing practices historically precede colonization and are kept alive through turmoil and legal restraints. Their preobjective stance situated in immediacy and indeterminacies is fluid yet strong, and may also explain the facility of the Rastafarian movement to travel and make itself at home in various locations. I hence prefer to find in the standard to avoid standardization a serious healing strategy enabling one to be well in-the-world. In these practices medicine is made to work in specific ways.

I have mainly learned about medicines and A. *afra* from *Xhosa izangoma* and Rastafarian *bossiedoktors* whose practices illustrate contrasts as well as similarities. The art of healing with *muthi* in Xhosa and Rastafarian practices lies in detachment from processes of standardization of everyday life. Both Xhosa and Rastafarian ways of being consider the 'inhabited world as sentient' (Ingold 2011: 12), a 'world as witty agent and actor' (Haraway 1988: 596). Like the world, people are in perpetual becoming. Most *isangoma* and Rastafarian healing strategies appear to rely upon ridding oneself of accumulated standards nested in one another to make way for new orders, even if these strategies differ between settings and individual healers. Rastafarians strive for this stance at all times through *livity*, while Xhosa healers reach this stance through intensified, phased performances. Rastafarian *livity* is knowing by feeling the basic unity of mankind, the shared human condition that unites all forms of life; it is a preontological consciousness to remain in the everyday as a way of being, very much in line with phenomenological approaches. It is about immediacy and personally experiencing divinity in the everyday as a means towards a better common future.

Rastafarians enact a preontological state of consciousness in the everyday as a means to heal the nation. The Rastafarian movement offers 'the opportunity to belong to a large collective grouping without the restrictions of formal rules, regulations, and bureaucracy' (Yawney 1978: 102). It is the great respect for life that grounds Rastafarians' refusal of all general organization, of any central power and of any dictator or chief. Prophets are sources of inspiration and creativity rather than authority. 'From this respect of life unfolds the fact that the movement structurally constitutes anarchy' (Constant 1982: 68). The standard of 'unstandardization' is maintained by keeping knowledge in the preontological everyday practices of reality, which are multiple and indeterminate. Xhosa *isangoma* expertise is in facilitating this movement of opening, precisely to rid the person to be healed and themselves of all forms of standardization, including language; repetitive rhythms, dance, sounds, glossolalia, all aim to reach a preontological state as a means to bring the (dis) eased person back into order within the world. This 'preobjective' state is prized by the healers and by those who work closely and explicitly with indeterminacies, connectivities and *muthi* multiplicities or 'medicine multiple' (Laplante forthcoming). Following the flow of doing medicine with Xhosa and Rastafarian healers led me to taste, smell, collect, listen, understand, transform, imagine, embody and feel the plants, people and place. These are the pathways to both healing and gaining the abilities to heal others through medicines, and these connections are crucial, and they are primarily done through immersion in sounds.

With the explicit histories and ontologies brought into being as South African indigenous medicine, it seems that *muthi* on trial through an RCT should be more about embodying sounds in life and less about isolated molecules (see Laplante 2009b), or at least more about their interconnections through lived bodily experiences. *Muthi* appears to be less about 'objects', such as plants, barks and minerals, than about the ways to relate accordingly through these mediums. Health and healing is not thought to be inherent in objects. The plant is rather a medium through which ancestral shades and spirits can be activated to affect one's life course in a beneficial manner. With the illegality of *isangoma* work since the eighteenth century, *muthi* has, however, more easily come to be equated with plants as objects that can be bought in the market and prepared conveniently to alleviate certain (dis)eases. If it is *isangoma* indigenous ways of healing that are on trial in the RCT, the process of singling out a plant, let alone a molecular configuration within that plant, is definitively not the way of recognizing these ways. To do so, understanding the efficacy of indigenous medicine should shift to understanding how bodily configurations can be transformed through diverse techniques and strategies and how skills can be acquired through the embodiment of sounds, dance and performance to modify the life course of both the initiate and the supporters in vivo.

As stated earlier, the RCT occurs through making the plant 'natural' as well as making human bodies into passive 'objects' upon which the effects of different 'natural' stimuli, such as a plant or a sound, are tested. Gouk (2005) explains how a shift of paradigm from 'nature as musical' to 'music is natural' was necessary to create the possibility to test the effects of music on humans: 'It was only after Newton succeeded in unifying the mathematical principles that underlay manifest mechanical actions and occult attractive forces in his new physics that the paradigm became "music is natural" and its effects on human nature became amenable to medical and scientific experiment' (ibid.: 104). As with music made natural, such is the case with medicine. In making medicine 'natural' and separating it from humans, it becomes possible to test the effects of medicine on humans. Of humans and medicines understood as separate entities, it is only the effects of the medicine upon human physiological processes that are of interest in an RCT; the effects of humans on, in or with medicine in mediums are of no interest.

Testing 'therapeutic sounds' in clinical trials today translates into clinical trials claiming to 'rehumanize' medicine in collaborations with musicians in order to understand the effects of music 'on' patients. Early experiments first invited the musicians into the clinics, but researchers afterwards decided that they rather needed to bring the 'patients' into the concert rooms, since this maximized therapeutic effects (Chémali 2010). Within this rationale, it is plausible to consider that music performed in the townships, by, for and with people living there, can act more intensively in listeners. What Chémali's trial, however, still overlooked is the patient's ability (agency) to hear receptively (or not), as well as the ability of the musician to communicate through performance. The 'rehumanized' clinical trial still does not account for what the 'patient' does to the music and the performance, nor can it account for such practices as 'doing *ngoma*' where there is no stage to separate the audience from the performance and all partake in it together. Further, such trials look solely at the effect of specific sound configurations on the neurobiological processes of a patient, notwithstanding their felt meanings in context. Should such a trial aim to understand efficacy in a *ngoma* drumming session, they would thus overlook the effects sounds have in the healers' body, sounds that enable them to connect with a plant and use it effectively to heal themselves and others. We are not close to being able to reinsert the healer into his accounts of the world, nor are we able to appreciate sounds, people and plants as connective and active healing agents. Further, since an expert is 'doing *ngoma*', one has to wonder why external 'verification' from distant observers would be needed at all; the very distance from the practice rather makes it unfeasible to follow what is going on.

Muthi on trial in my study is even farther away from 'indigenous knowledge', since the RCT is solely preoccupied with a single plant, a plant that *izangoma* and *izinyanga* refuse to use in their practices since they disagree with the way it was cultivated, explaining that it has lost its life, or efficacy, in its farmed form. The plants to be used in healing are also engaged through dreams and visions indicating their location. That the *izangoma* agree with the *izinyanga* on this point may have to do with a shared ontology based on sounds and movement, as well as shared hopes. Perhaps this similarity also lies in the importance given to the intention and ability of the person who tends, prepares and collects the plant. This positioning emphasizes a particular shared way of 'being with' that does not dissociate humans from 'nature', instead relying on these entanglements in order for the plant to 'work'.

Not only is the plant singled out, but so is disease. Health is *livity* for Rastafarian *bossiedoktors* and connectivity for *izangoma*. The aim is respectively expressed as healing the nation through people and healing people in the world. This finding corresponds to what Devisch describes with regards to Yaka 'cults' in Zaire:

> Illness and healing are multilayered realities. Contrary to current biomedical notions that view health merely as the absence of organic dysfunction, the members of Yaka cult and self-help groups in Kinshasa and the villages appear to interpret health and individual well-being as resulting from specific relations seen in a much broader context. To be in good health depends on the relations between people – or between the individual, the group, and the environment or life-world, and results from the vital integration of elements which also determine the fertility of the social group, success at school or work, and the moral and material well-being or continuity of the family group. Being in good health therefore means being whole, that is being integrated in meaningful ways into the relational fields of body, group, and life-world. (Devisch 1993: 31)

This is not under trial; rather, it is assumed guilty of introducing bias before given any chance of proving its innocence, let alone its usefulness and efficacy. The disease that the preclinical trial of *A. afra* aims to eradicate is one that is embedded within disrupted living conditions, poverty and malnutrition, namely precisely what is dealt with by the healers with, through and beyond the plant being tested. By attending to injustice, inequality and loss of direction in life, healers take the toil and turmoil of life upon themselves and enter into more intricate relations within mediums in the inhabited world in ways to embellish it. As such, I have begun to explore my own hopes for research, a way forwards I explore for anthropology.

Ethics or Aesthetics?

The RCT gold standard is a scientific research device, model, practice and ethical procedure that holds the promise of closure through fact-finding to ensure the 'safety' and 'efficacy' of medicine. As in scientific fact-finding, 'in ethics, the promise of closure, or at least temporary consensus, through reasoning is widely shared' (Mol 2002: 177). In an attempt to understand this promise, I've found it useful to delve into practices in preparation to conduct an RCT, finding them to open to politics, even at the molecular level. 'The term politics resonates openness, indeterminacy' (ibid.). The 'common good' the RCT puts forth is thus inevitably multiple as well as challenged by other strong politics hoping for a different kind of common world. The 'good', the paths organizing engagements in medicine in the preclinical of *A. afra,* were found to flow through hopes in indigeneity as well as in scientific innovation, both motions and middle paths attempted by following the same design. The RCT is a research device of a legislative kind, made of rules, protocols, procedures and standards, with distant authorities of all sorts pulling practices in one direction to follow preestablished guidelines. This line, however, bends and pulls from the middle in numerous other directions throughout the process.

In the RCT's travels, its practices become less 'taken for granted' because it is uprooted as well as challenged by other ways of organizing experience in the world. It also becomes less 'ethical' in relation to the context at hand, particularly in its dealings with indigenous medicine, which works through aesthetics in therapeutics. I have thus similarly been concerned with worthwhile ways of living with medicine in the everyday and expert practices, namely, with the kinds of life that are of concern and that guide the efficacies in doing medicine. I have in particular been interested in the healing efficacies of sound, as they both embellish and heal in the everyday. In this move, I may have come to an ethics of an aesthetic kind. Not one that is normative, but one that is lived aesthetically or poetically in its immediacy and intricacies in the world – by doing lived research immersed in medicine multiple, for instance, taking action during research by attending to and partaking in worthwhile ways of living with the real in felt worlds, through mediums, lives, people, plants and things interlaced.

Ethics in research is a particular engagement in-the-world; it can be of the legislative kind (RCT), the critical kind (proposing new legislation), the mingling kind (leading to new ethics that might be taken up by the critical kind) or it can become of the normative kind, as Mol (2002: 152–57) suggests we describe styles or ways of attending to methods. I here want to explore the idea of aesthetic sensibilities in research as a kind of ethics for anthropology. I suggest ways to engage with people and things in research can make worthwhile ways of living with the real. We have somewhat adjusted to this in anthropology. If, for instance, asking a healer to sign a consent form breaks with meaningful engagements, we agree in anthropology to follow other routes. I suggest we take this flexibility a step further. Lived research attuned with aesthetics sensibilities can be cultivated. With aesthetic sensibilities I thus allude to a reawakening of the senses in engaging with medicine, creating and appreciating beauty in making medicine and, correspondingly, in healing and being healed.

Such research is one of engagement in the world, steered towards doing worthwhile ways of living with the real as opposed to solely being concerned with collecting data for future analysis, as also strongly suggested by Ingold (2011, 2013). This implies investing mind and body, deepening relations in the world, as Xhosa and Rastafarian healers do, which inspired me to follow a new approach to grasp what was going on in a preclinical trial; in doing so I became immersed in the ways of living with the real in this process, as well as in what it feels like to make medicine work in the peripheries of this process. This approach necessarily led to questions about ways of being human as well as how new technological innovations might transform these ways. I thus argue aesthetics can be a concern in the very process of doing medicine, engaging in the process with people and things not as objects to describe or manipulate but as kin sharing a common world, in particular when dealing with live plants, but just as much when delving into ways of orienting human lives at the molecular level and beyond.

It is precisely by inhabiting matter that we can find new forms of aesthetic sensibilities. To do so we may need to let wither away a positioning of mastery of the environment, of control over life, and rather inhabit life in all of its indeterminacies. Perhaps, following Montaigne, 'we must realize there is no intellectual and moral supremacy of humans on animals since one and the other are submitted to the same natural constraints' (Descola 2005: 245, my translation). The same can apply with regard to plants, although I would argue in both cases that it is not submission to the same 'natural constraints' that unites humans and nonhumans, but openings to similar environmental potentialities to develop skills and capacities of engagement. If, as Küchler (2006) suggests, technological innovations are heading towards intelligent materials transcending the mind-body, nature-culture dichotomies, transforming meaning into matter, we should do the same in anthropology and follow the rhythm of these innovations or, as I have proposed, follow the rhythm of improvisations in-the-world beyond what is achieved within controlled environments. In practice, this turns a desire to 'help' or 'save lives' from an external standpoint towards a deeper engagement with humans and nonhumans in finding good ways of carrying on within life-making processes, perhaps including proteins and genes, memories and desires, poetics and aesthetics as lived and felt.

Life may be at the centre of this composition, not a hierarchical life divided between its biology and its humanity, but life in all of its entanglements. Current practices enacted in the RCT are forms of enquiry into the processes of life. They have, however, turned upon themselves to solely demonstrate the correctness of their method rather than attend therapeutics. With the emphasis brought to following ethical guidelines, the RCT has become one of the greatest parts of current ethical policies. In current ethical policies, usually only a short section is allocated to inductive methods, which leave open possibilities to learn from ongoing practices; prior policing also inhibits these approaches from freely engaging in the world to learn. The anxiety generated by a distant third party might not make for more ethical engagements in life; it might even do the contrary, since the researcher feels constrained and monitored rather than fully engaged. A dissociated way of doing research has repercussions on the kinds of technologies made as well as on the ways of living with the real both inside and outside the experiment and laboratory. It is thus these spaces that need to be recomposed for any other forms of engagements in medicine to emerge from these realms.

My proposal is to embed oneself in life during the research process as humans sharing mediums with other forms of life, and not as externalized observers. Distant 'objective' forms of knowledge manage uncertainty in very restricted manners, attending solely to the measurable and the representational. These forms of objectivities are convenient for the mass production of technologies. Making parts of the world 'natural' is convenient in transforming the world into bioresources suitable to testing in humans. No ethical clearance is currently needed to manipulate plants in one way or another. These are conditions of possibility placing experts above or far from what they aim to know in plants for medicine. They lead to particular ways of interfering with life in a broader sense without attending to these issues directly.

Lived research is also a form of life and a way to interfere with life, as well as a way to manage uncertainty in proximity. Research in life can diminish erroneous interpretations of people's behaviours by investigating their intentions. Similarly, proximity with plants can develop ways for nonhumans (plants, for instance) and humans to live together in beneficial ways, other than through hierarchical relationships of control, regulation and manipulation. Aesthetic sensibilities can slow down the pace of production of technologies in delving into these intricacies, yet this pace is already bottlenecked downstream in bureaucracies and ethics committees. Improvisation in medicine is usually bottlenecked for a number of administrative and financial reasons, in keeping with protocols, procedures and legislation. The translational validation model of reversed pharmacology brought forwards by TICIPS researchers is precisely an initiative to avoid such bottlenecks downstream, in taking more time to learn in life with regard to clinical feasibility before undergoing laboratory procedures. These obstacles are unavoidable unless we adopt ways of doing technologies in the flow of life. Developing aesthetic sensibilities in situ and in vivo is necessary so that scientific experts can work together with healers and indigenous forms of knowledge. Doing so could address the RCT's current blind spots as well as dissolve the model by opening it up to very different kinds of possibilities of living with the real.

This is the kind of research in which anthropology excels, even if anthropology also often slips outside of the movements in life when it attempts to reproduce a given model or when it sets 'data' or 'ethnography' as an objective set of information to collect and to analyze, even if these data are most often anchored and obtained in the phenomenological experience in the world. Anthropology's legitimacy, as that of indigenous medicine, however, remains secondary even today, perhaps since both attend to greater forms of life, beyond life itself, as well as attending to the visible, the lived and the felt. Before the mid-nineteenth century, aesthetics in research were openly bounded to 'objectivity', with scientists working closely with artists. The moralization of objectivity turned to the policing of artists at the end of the nineteenth century, followed by the policing of the scientists' temptations themselves. Self-effacement became a moral virtue for scientists. The search for this rendition of objective representation was a moral as much as a technical quest (Daston and Galison 1992: 117). It has not left positivism, and the RCT's continuing existence demonstrates its becoming a 'way of life', although one critiqued by most of the scientists involved, as it does not fit with their practices. Vigilance against the temptations of theorizing, aestheticizing and pouring evidence into preconceived moulds remains the image of 'rigorous' scientific research today.

Something else has occurred more recently, perhaps even enhancing this disengagement of the scientist in the world. It seems that self-effacement and self-restraint has moved even further from the scientists themselves and into the hands of a proliferation of ethical committees that Rose (2007) names the new pastorate: clinical trial monitors, international review boards and university ethical policies all add up to ensure ethics enacted in the RCT remain ever more distant from the researcher, who must already be distant from his own self. For Rose, new forms of pastoral powers are forms of 'biomorality' or 'somatic ethics' that bring the body and the 'bio' to the centre of our forms of life (ibid.: 256). These pastoral powers take shape in and around our genetics and our biology, entangling research scientists, biotechnology executives, genetic counsellors in ethics and ethopolitics (ibid.: 254). A somatic ethic is also found to be taking shape in nongovernmental organizations (NGOs), philanthropy and humanitarian efforts addressing problems of the poor, of neglected disease and access to medicine. Rose suggests the 'biological reductionism' upon which they proceed, with concerns for 'life itself', should be the grounds for a certain optimism rather than the basis of critique (ibid.: 255). I have found this to be problematic, perhaps because I have stepped outside these elite circles and into the worlds of the 'poor'. I tend to agree with Rose's notion of 'somatic ethics', or 'ethics as a way of understanding, fashioning, and managing ourselves in the everyday conduct of our lives' (ibid.: 257), which draws on Gilles Deleuze and Michel Foucault's technologies of the self. I, however, remain critical of reducing concerns to 'life itself', since I find this is the very root of the problem, which thus needs to be problematized rather than dealt with. It also in my view is intimately linked to the issue of doing research from an externalized standpoint. Scientific research currently adheres to ethics as a code of moral obligations, a distant preexisting model that is not always easy to enact, fashion and manage in practice, nor is it necessarily felt as meaningful.

According to Ali and Pandian (2009), ethics in the humanism of the enlightenment have 'undone' the prehumanist practices such as ascetics, techniques of the self, etiquette and rhetoric. With the triumph of Kantian humanism, 'ethical' thought became ideal, universal and rational, therefore making it amenable to application to all human beings. In this form the ethical is made into an a priori design to follow; universal moral imperatives are assumed inherent to human rationality itself, strangely resembling the RCT. Kantian ethics unifying all humans under the same rights and values resolved important issues of equity in life in its broader sense, but through time it also translated into a real attenuation of ethical preoccupations in scientific practice; it practically loses all relevance because of its idealist, distant nature, becoming a bureaucratic act and a constraint dissociated from the felt goodness of the act. Formal ethical guidelines outliving their moments of emergence become a separate domain of their own, a form of external ethical policing. The RCT procedure was described in this very way by most of the scientists involved in the preclinical trial I followed.

Ali and Pandian (2009) propose to broaden the current field of ethics and morals. In reviewing the varying genealogies of virtue in

Asia, they envision ethics as a reflexive practice of the 'liberty' between the known and the given, according to Foucault's (1997) expression, similarly to what Rose (2007) suggests with somatic ethics,
but beyond expert committees and governmental agencies as well. In
this perspective, ethics lead back to disparate conceptions, to codes
of moral obligations, to assemblages of habits and customs, to ex-
ercises of self-discipline, to ways of living well with and for others,
to domains favourable for the exercise of aestheticism, skills and
virtue, to places of requirements on behalf of others (Ali and Pandian 2009: 44–45). Ali and Pandian (ibid.) suggest leaving aside the
normative aspects of ethics, those of cultural specificities of moral
traditions and ethical practices, and instead paying attention to the
connections between the conceptual articulation and the everyday
practices, and to the concretization of these acts in the body of experience. Without an emphasis on embodied knowledge, reflexivities,
bodies and senses through the act of research, I would not have been
able to grasp both the RCT and indigenous healing practices. Leaving ethics open and flexible to feel ways of living with the real might
create a favourable terrain for the proliferation of new and inventive
aesthetics and worthwhile ways of living with the real and thus for
doing medicine in more meaningful ways.

Research in anthropology is a reflexive practice in proximity with
people and things in the world. It can pay attention to the measurable, refraining from thinking what it might feel like to live with
technologies, and it can, as with the healers, deepen feelings in the
world to transform ways of living with the real. Healing is an art
and a science; its tools must fit certain aesthetics as well as agree
with the moral values of those cared for to be useful. Research in
anthropology is also an art and a science, one that can also heal issues in life if done successfully together with the people it concerns.
Anthropological practices will not solve everything, but they may be
part of real solutions to real problems in attending to entanglements
in life. The emergence of bioethics may be useful, but they also do
not fully problematize life itself; inversely, current ethics are primarily concerned with a humanity dissociated from its biology and from
other living beings. This divide might also contribute to making ethics difficult to cope with in practice, in combination with its externalization from the world.

'The irrelevance of ethics can be seen when considering universal ethical formulations of justice and equity that do not begin with the local moral conditions of poor people, those experiencing the systematic injustice of higher disease rates and fewer health-care resources because of their positioning at the bottom of local social structures of power' (Kleinman 1999: 72). The irrelevance of ethics is felt when it is imposed as a moral code from above and beyond possibilities of research methods in the world or the context at hand, which requires flexibility rather than closed grids and models. This applies to RCTs, which are the currently endorsed 'best ethical procedures' to follow. RCTs require new flexibilities to really become 'ethical' in the sense of being able to fit meaningfully within mediums. 'For almost all of us, everyday life experience in communities and networks – no matter how influenced we are by global forces of communication, commerce, and the flow of people – centers on "what is locally at stake"' (ibid.: 70). 'What is locally at stake' for scientists involved in the RCT has become following complex ethical procedures and a method that sometimes take precedence over what is locally at stake for the people concerned; ways of manipulating life are often left unattended, as is imagining how following procedures might make particular kinds of life, which will become part of making our common world.

Following how people and things interweave in and around an RCT offers insight into how to make room for more meaningful routes, routes that are not caught in a sole predesigned web, but are part of motions in meshworks. The dominant way of doing research and establishing rules is deductive rather than inductive. The assumption that man-made models of nature are the most convenient to conduct objective experiments and make the world 'fair' and 'safe' presides. Should they be dominant, inductive approaches would not require providing such direction 'out of context', but would rather follow research through the concrete dilemmas encountered. Research through proximity and in context assures the management of uncertainty, since people can share their interpretations and are involved in the written account, the script or the performance. Within the formal infrastructure and current biopolitical economy of global health, encounters between scientists and healers occur on the surfaces. The ways those trials currently unfold are restrictive and perhaps harmful and demeaning to human dignity, even with a will of 'recognition'. How to move beyond this impasse seems to point towards the need to do research differently.

Perhaps we need to leave the old anthropocentric (human-centred) values to ground ourselves in ecocentric (earth-centred) values, as suggested by Capra (1996: 11), following what has been learned in molecular biology, requiring that all living beings be considered members of ecological communities bound together. 'When this deep ecological perception becomes part of our daily awareness, a radically new system of ethics emerges. Such a deep ecological ethics is urgently needed today, and especially in science, since most of what scientists do is not life-furthering and life-preserving but life-destroying' (ibid.). Ingold finds that the current attitude is best described as anthropocircumferentialism (2000: 218) in the sense that humans are not seen as part of the world but outside of it, or surrounding it. In line with this, he proposes that a truer anthropocentrism would bring humans in the world, surrounded by the world. Current anthropocircumferentialism presupposes a global perspective, one outside the world, such as the perspective taken when one aims to 'save' and 'protect' life and the environment from the outside rather than engage with it in more beneficial ways.

Thus, living things are classified and compared, and their kinds enumerated, in terms of intrinsic properties that they are deemed to possess by virtue of genealogical connection, irrespective of their positioning in relation to one another in an environment. This is the basis for the modern concept of biodiversity. As far as human differences are concerned, these are typically understood in terms of a concept of cultural diversity, that is, seen as analogous to biodiversity rather than as an extension of it. A humanity that is common to all is superimposed on this all-embracing environment, which is 'profoundly alien to human experience' (ibid.: 217).

It is human as well as nonhuman experience that weaves the world together and that should be the null point of ethics. Tailoring the senses to grasp phenomena as they emerge and as we engage with them becomes a route to recompose our world as inhabitants in-the-world. In other words, living, performing in ways to make this world more aesthetic and meaningful, including during research, might inspire others to do so as well in the everyday. Aesthetics assures meaning is weaved into the world. It also seems to be a fruitful way to manage uncertainty in engaging in medicine.

Attuning aesthetic sensitivities is also a journey and a form of knowledge that can make life as well as modify life. My journey has been to follow traces of A. *afra* in a process to make it into a bio-pharmaceutical. It has taken me into laboratories, townships and valleys, through different ways of organizing experience in the world, including Rastafarian, Xhosa and scientific roots, as well as in the trails they currently follow. All are in the world rather than in its periphery, some aiming for deeper engagements in the world, others (dis)engagements, to best know and modify life. Political economies of hope weave themselves through these practices, bringing natures and cultures to life in varying ways, sometimes ethically, other times aesthetically. I foresee the need for greater enhancement of aesthetic sensibilities in doing anthropology.

✣ References

Adams, V. 2002. 'Randomized Controlled Crime: Postcolonial Sciences in Alternative Medicine Research', *Social Studies of Science* 32(5–6): 659–90.

Adams, V., et al. 2005. 'The Challenge of Cross-Cultural Clinical Trials Research: Case Report from the Tibetan Autonomous Region People's Republic of China', *Medical Anthropology Quarterly* 19(3): 267–89.

Agamben, G. 1997. *Homo sacer: Le pouvoir souverain et la vie nue.* Paris: Éditions du Seuil.

———. 2006. *L'Ouvert: De l'homme et de l'animal.* Paris: Rivages poche / Petite Bibliothèque.

Agrawal, A. 1995. 'Dismantling the Divide Between Indigenous and Scientific Knowledge', *Development and Change* 26(3): 413–39.

———. 2002. 'Indigenous Knowledge and the Politics of Classification', *International Social Science Journal* 54(173): 287–97.

Ali, D. and A. Pandian. 2009. 'Généalogies de la vertu: Pratiques éthiques en Asie du Sud', *Anthropologie et Sociétés* 33(3): 43–60.

Alloa, E. 2008. *La résistance du sensible: Merleau-Ponty—critique de la transparence.* Paris: Editions Kimé.

Arendt, H. 1958. *The Human Condition.* Chicago: University of Chicago Press.

Ashforth, A. 2000. *Madumo: A Man Bewitched.* Chicago: University of Chicago Press.

Avula, B., et al. 2009. 'Quantitative Determination of Flavonoids by Column High-Performance Liquid Chromatography with Mass Spectrometry and Ultraviolet Absorption Detection in Artemisia afra and Comparative Studies with Various Species of Artemisia Plants', *Journal of AOAC International* 92(2): 633–44.

Backster, C. 1968. 'Evidence of a Primary Perception in Plant Life', *International Journal of Parapsychology* 10(4): 329–48.

———. 1973. 'Evidence of a Primary Perception at Cellular Level in Plant and Animal Life', unpublished manuscript. Backster Research Foundation, Inc.

Balée, W. 1994. *Footprints of the Forest: Ka'apor Ethnobotany—the Historical Ecology of Plant Utilization by an Amazonian People.* New York: Columbia University Press.

Bannerman, R., J. Burton and W.C. Ch'en. 1983. *Traditional Medicine and Health Care Coverage: A Reader for Health Administrators and Practitioners.* Geneva: World Health Organization.

Beaucage, P., et al. 1997. 'Le savoir ethnopharmacologique des Nahuas de la Sierra Norte de Puebla (Mexique): structure et variation', *Recherches Amérindiennes au Québec* 27(3–4): 19–30.

Beck, U., A. Giddens and S. Lash. 1994. *Reflexive Modernization: Politics, Tradition and Aesthetics in the Modern Social Order.* Stanford, CA: Stanford University Press.

Becker, H.S. 1963. *Outsiders: Studies in the Sociology of Deviance.* New York: Free Press.

Behrens, K.G. 2013. 'Towards an Indigenous African Bioethics', *South African Journal of Bioethics and Law* 6(1): 32–35.

Bernard, C. 1865. *Introduction à l'étude de la médecine expérimentale.* Paris: Garnier-Flammarion.

Birke, L. 2012. 'Animal Bodies in the Production of Scientific Knowledge: Modelling Medicine', *Body & Society* 18: 156–78.

Black, J. 1992. *Arrow of the Blue-Skinned God: Retracing the Ramayana through India.* Boston: Houghton Mifflin.

Bourdieu, P. 1977. *Outline of a Theory of Practice,* trans. R. Nice. Cambridge: Cambridge University Press.

———. 1984. *Distinction,* trans. R. Nice. Cambridge, MA: Harvard University Press.

Bourdieu, P. and L.J.D. Wacquant. 1992. *Réponses: Pour une anthropologie réflexive.* Paris: Seuil.

Bowker, G.C. and S.L. Star. 2000. *Sorting Things Out: Classification and its Consequences.* Cambridge, MA: MIT Press.

Brives, C. 2013. 'Identifying Ontologies in a Clinical Trial', *Social Studies of Science* 43(3): 397–416.

Bryant, A.T. 1966. *Zulu Medicine and Medicine-Men.* Cape Town: C. Struik. First published 1909 by Annals of the Natal Museum.

Canguilhem, G. 2008. *Knowledge of Life,* trans. S. Geroulanos and D. Ginsburg. New York: Fordham University Press.

Capra, F. 1996. *The Web of Life: A New Scientific Understanding of Living Systems.* New York: Anchor Books.

Carrier, N., J. Laplante and J. Bruneau. 2005. 'Exploring the Contingent Reality of Biomedicine: Hepatitis C, Injecting Drug Users and Risk', *Health, Risk and Society* 7(2): 123–40.

Cartwright, N. 2007. 'Are RCTs the Gold Standard?', *BioSocieties* 2(1): 11–20.

Chao, C.-F. 2009. 'Dynamic Embodiment: The Transformation and Progression of Cultural Beings through Dancing', *Taiwan Journal of Anthropology* 7(2): 13–48.

Chapin, M. 2004. 'A Challenge to Conservationists', *World Watch Magazine,* November/December 17–31.

Charlton, K.E., et al. 1994. *Food Habits, Dietary Intake and Health of Elderly Coloureds in the Cape Peninsula.* Cape Town: HSRC/UCT Centre for Gerontology.

Check, E. 2007. 'After Decades of Drought, New Drug Possibilities Flood TB Pipeline', *Nature Medicine* 13(3): 266.

Chémali, K.R. 2010. 'The Science of Music Rehumanizing Medicine: Scientists and Musicians Discover the Importance of Their Colloboration', *Music and Medicine* 2(2): 73–77.

Chernoff, J.M. 1979. *African Rhythm and African Sensibility*. Chicago and London: University of Chicago Press.

Claassen, J. 2005. 'The Gold Standard: Not a Golden Standard', *British Medical Journal* 330(7500): 1121.

Classen, C. 1999. 'McLuhan in the Rainforest: The Sensory Worlds of Oral Cultures', in D. Howes (ed.), *Empire of the Senses: The Sensual Culture Reader*. Oxford: Berg, pp. 147–163.

Classen, C., D. Howes and A. Synnot. 1994. *Aroma: The Cultural History of Smell*. New York: Routledge.

Clément, C. 2009. 'Claude Lévi-Strauss, une vie', *L'Express*, 3 November.

Clifford, J. and G.E. Marcus. 1986. *Writing: The Poetics and Politics of Ethnography*. Berkeley: University of California Press.

Collot, P.-A. 2007. 'La protection des savoirs traditionnels, du droit international de la propriété intellectuelle au système de protection sui generis', *Droit et Cultures* 53: 181–209.

Constant, D. 1982. *Aux sources du reggae: Musique, société et politique en Jamaïque*. Paris: Parenthèses.

Coreil, J. 1990. 'The Evolution of Anthropology in International Health', in J. Coreil and J.D. Mull (eds), *Anthropology and Primary Health Care*. Boulder, CO: Westview Press, pp. 3–27.

Crampton, H. 2006. *The Sunburnt Queen*. London and San Francisco: Saqi Books.

Crouzel, I. 2000. 'La "Renaissance Africaine": Un discours sud-africain?', *Politique africaine* 77: 171–82.

Csordas, T.J. 1988. 'Elements of Charismatic Persuasion and Healing', *Medical Anthropology Quarterly* 2: 121–42.

———. 1990. 'Embodiment as a Paradigm for Anthropology', *ETHOS* 18: 5–47.

———. 1996. 'Imaginal Performance and Memory in Ritual Healing', in C. Laderman and M. Roseman (eds), *The Performance of Healing*. New York: Routledge, pp. 91–114.

———. 2007. 'Transmutation of Sensibilities: Empathy, Intuition, Revelation', in A. McLean and A. Leibing (eds), *The Shadow Side of Fieldwork: Exploring the Blurred Borders between Ethnography and Life*. Malden, MA: Blackwell, pp. 106–16.

Curry, A. 1968. 'Drugs in Rock and Jazz Music', *Clinical Toxicology* 1(2): 235–44.

Daniels, M. and A.B. Hill. 1952. 'Chemotherapy of Pulmonary Tuberculosis in Young Adults: An Analysis of the Combined Results of Three Medical Research Council Trials', *British Medical Journal* 1(4769): 1162–68.

Das, V. 2003. 'Technologies of Self: Poverty and Health in an Urban Setting', in *Shaping Technologies, Sarai Reader 03*, Delhi: Autonomedia. pp. 95–103.

Daston, L. and P. Galison. 1992. 'Objectivity and the Escape from Perspective', *Social Studies of Science* 22(4): 597–618.

De Craen, A.J.M., et al. 1999. 'Placebos and Placebo Effects in Medicine: Historical Overview', *Journal of the Royal Society of Medicine* 92(51): 1–515.

Deleuze, G. and F. Guattari. 1980. *Mille Plateaux: Capitalisme et schizophrénie*. Paris: Éditions de Minuit.

Descola, P. 2005. *Par-delà nature et culture*. Paris: Gallimard.

Desjarlais, R.R. 1992. *Body and Emotion: The Aesthetics of Illness and Healing in the Nepal Himalayas*. Philadelphia: University of Pennsylvania Press.

Desjarlais, R.R. and J. Throop. 2011. 'Phenomenological Approaches in Anthropology', *Annual Review of Anthropology* 40: 87–102.

Devisch, R. 1993. *Weaving the Threads of Life: The Khita Gyn-Eco-Logical Healing Cult among the Yaka*. Chicago: University of Chicago Press.

Diehl, H.S., A.B. Baker and D.W. Cowan. 1938. 'Cold Vaccines: An Evaluation Based on a Controlled Study', *JAMA* 111: 1168–73.

Douglas, M. 1973. *Natural Symbols: Explorations in Cosmology*. New York: Vintage Books.

Dreyfus, H. and P. Rabinow. 1984. *Michel Foucault: Un parcours philosophique*. Paris: Gallimard.

Duggan, P.F. 1992. 'Time to Abolish "Gold Standard."' *British Medical Journal* 304: 1568–69.

Dugmore, H. and B.E. van Wyk. 2008. *Muthi and Myths from the African Bush*. Pretoria: Marula Books.

Dykman, E.J. 1908. *De Suid Afrikaanse Kook-, Koek- en Resepte Boek*, 14th improved impression. Paarl (Cape Colony): Paarl Printers.

Dyson, A. 1998. *Discovering Indigenous Healing Plants of the Herb and Fragrance Gardens at Kirstenbosch National Botanical Garden*. Cape Town: National Botanical Institute Printing Press.

Edmonds, E.B. 2003. *Rastafari: From Outcasts to Culture Bearers*. Oxford: Oxford University Press.

Ellen, R. and H. Harris. 2000. 'Introduction', in R. Ellen, P. Parkes and A. Bicker (eds), *Indigenous Environmental Knowledge and its Transformations: Critical Anthropological Perspectives*. Netherlands: Harwood Academic Publishers, pp. 1–34.

Epstein, S. 2007. *Inclusion: The Politics of Difference in Medical Research*. Chicago: University of Chicago Press.

Erlmann, V. 2005. *Hearing Cultures: Essays on Sound, Listening, and Modernity*. Oxford: Berg.

Erlund, I. 2002. 'Chemical Analysis and Pharmacokinetics of the Flavonoids Quercetin, Hesperetin and Naringenin in Humans', Ph.D. dissertation. Helsinki: University of Helsinki.

Escobar, A. 1984–85. 'Discourse and Power in Development: Michel Foucault and the Relevance of his Work to the Third World', *Alternatives* 10: 377–400.

Eskrine, N.L. 2005. *From Garvey to Marley: Rastafari Theology*. Gainesville: University Press of Florida.

Etkin, N.L. 1990. 'Ethnopharmacology: Biological and Behavioral perspectives in the Study of Indigenous Medicines', in T. Johnson and C.F. Sargent (eds),

Medical Anthropology: A Handbook of Theory and Method. New York: Prae-
ger, pp. 149–58.

———. 1993. 'Anthropological Methods in Ethnopharmacology', *Journal of
Ethnopharmacology* 38(17): 93–104.

Evans, J.G. 2010. 'East Goes West: *Ginkgo biloba* and Dementia', in E. Hsu
and S. Harris (eds), *Plants, Health and Healing: On the Interface of Ethnobotany and Medical Anthropology.* Oxford and New York: Berghahn Books,
pp. 229–61.

Evans-Pritchard, E.E. 1937. *Witchcraft, Oracles and Magic Among the Azande.*
Oxford: Clarendon Press.

Fachner, J. 2006. 'An Ethno-Methodological Approach to Cannabis and Music
Perception, with EEG Brain Mapping in a Naturalistic Setting', *Anthropology of Consciousness* 17(2): 78–103.

Fainzang, S. 1989. *Pour une anthropologie de la maladie en France: Un regard
africaniste.* Paris: Cahiers de L'homme.

Farnell, B. and C.R. Varela 2008. 'The Second Somatic Revolution', *Journal for
the Theory of Social Behaviour* 38(3): 215–40.

Farquhar, J. 1994. *Knowing Practice: The Clinical Encounter of Chinese Medicine.* Boulder, CO: Westview Press.

Fassin, D. 2000. 'Entre politiques de la vie et politiques du vivant. Pour une
anthropologie de la santé', *Anthropologie et sociétés* 24(1): 95–116.

———. 2006a. 'La biopolitique n'est pas une politique de la vie', *Sociologie et
sociétés* 38(2): 35–48.

———. 2006b. *Quand les corps se souviennent: Expériences et politiques du sida
en Afrique du Sud.* Paris: La Découverte, pp. 202–70.

———. 2010. 'Ethics of Survival: A Democratic Approach to the Politics of
Life', *Humanity* 1(1): 81–95.

Favret-Saada, J. 1977. *Les mots, la mort, les sorts.* Paris: Gallimard.

Ferguson, J. 2006. *Global Shadows: Africa in the Neoliberal World Order.*
Durham, NC and London: Duke University Press.

Ferreira, M. 1987. 'Medicinal Use of Indigenous Plants by Elderly Coloureds:
A Sociological Study of Folk Medicine', *South African Journal of Sociology*
18(4): 139–43.

Ferreira, M., K. Charlton and L. Impey. 1996. 'Traditional Medicinal Use of In-
digenous Plants by Older Coloureds in the Western Cape', in H. Norman, I.
Snyman and M. Cohen (eds), *Indigenous Knowledge and its Uses in Southern
Africa.* Cape Town: HRSC, pp. 87-108.

Flint, K. 2001. 'Competition, Race, and Professionalization: African Healers
and White Medical Practitioners in Natal, South Africa in the Early Twentieth Century', *Social History of Medicine* 14(2): 199–221.

———. 2006. 'Indian–African Encounters: Polyculturalism and African Therapeutics in Natal, South Africa, 1886–1950s', *Journal of Southern African
Studies* 32(2): 367–87.

———. 2008. *Healing Traditions: African Medicine, Cultural Exchange, and
Competition in South Africa, 1820–1948.* South Africa: University of Kwa-
Zulu-Natal Press.

Foucault, M. 1976. *Histoire de la sexualité*, vol. 1, *La volonté de savoir*. Paris: Gallimard.

———. 1988. 'Technologies of the Self', in L.H. Martin, H. Gutman and P.H. Hutton (eds), *Technologies of the Self: A Seminar with Michel Foucault*. Amherst: The University of Massachusetts Press, pp. 16–49.

———. 1997. 'Polemics, Politics and Problematizations', in P. Rabinow (ed.), *Ethics: Subjectivity and Truth*. London: Allen Lane, pp. 111–19.

Friedman, L.M., C.D. Furberg and D.L. DeMets. 1998. *Fundamentals of Clinical Trials*, 3rd ed. New York: Springer-Verlag.

Gagneux, S. and P.M. Small. 2007. 'Global Phylogeography of *Mycobacterium tuberculosis* and Implications for Tuberculosis Product Development', *Lancet Infectious Diseases* 7: 328–37.

Geissler, P.W. and R. Prince. 2010. 'Persons, Plants and Relations: Treating Childhood Illness in a Western Kenyan Village', in E. Hsu and S. Harris (eds), *Plants, Health and Healing: On the Interface of Ethnobotany and Medical Anthropology*. Oxford and New York: Berghahn Books, pp. 179–224.

Geurts, K.L. 2002. *Culture and the Senses: Embodiment, Identity, and Well-Being in an African Community*. Berkeley: University of California Press.

Gibbs, W.W. 2003. 'The Unseen Genome: Gems among the Junk', *Scientific American* 289: 48–53.

Gibson, D. 2010. 'Negotiating the Search for Diagnosis and Healing Tuberculosis in Namibia. A Case Study of a Ju/'hoansi speaking man', *African Sociological Review* 14(2): 47–61.

———. 2011. 'Ambiguities in the Making of an African Medicine: clinical trials of *Sutherlandia frutenscens (L.) R.Br (Lessertia frutescens)'*, *African Sociological Review* 15(1): 124–37.

Gibson, J.J. 1986. *The Ecological Approach to Visual Perception*. Hillsdale, NJ: Lawrence Erlbaum.

Giddens, A. 1994. 'Living in a Post-Traditional Society', in U. Beck et al. (eds), *Reflexive Modernization: Politics, Tradition and Aesthetics in the Modern Social Order*. Stanford, CA: Stanford University Press, pp. 56–109.

Goldstein, K. 1995. *The Organism: A Holistic Approach to Biology Derived from Pathological Data in Man*. New York: Zone Books.

Good, B.J. 1994. *Medicine, Rationality, and Experience: An Anthropological Perspective*. Cambridge: Cambridge University Press.

Goodson, J.A. 1922. 'The Constituents of the Flowering Tops of *Artemisia afra*, Jacq.', *Biochemical Journal* 16: 489–93.

Gouk, P. 2005. 'Raising Spirits and Restoring Souls: Early Modern Medical Explanations for Music's Effects', in E. Veit (ed.), *Hearing Cultures: Essays on Sound, Listening and Modernity*. Oxford: Berg, pp. 87–106.

Gratton, D. 1986. 'Jah know: Étude sur le mouvement rastafari, l'usage du ganja et la famille en Jamaïque', MA dissertation. Montréal: Université de Montréal.

Green, L.J. 2008. 'Indigenous Knowledge', part 2, in N. Shepherd and S. Robins (eds), *New South African Keywords*. Ohio: Ohio University Press, pp. 132–42.

———. 2009. 'Anthropologies of Knowledge and South Africa's Indigenous Knowledge Systems Policy', *Anthropology Southern Africa* 31(1–2): 48–57.

Gruenwald, J. 2005. *Hoodia: Business Opportunity or Dangerous Business*. Retreived 13 November 2013 from http://www.nutraceuticalsworld.com/issues/2005-09/view_columns/eurotrends-hoodia-business-opportunity-or-dangerou/.

Hahn, R.A. 1995. *Sickness and Healing: An Anthropological Perspective*. London: Yale University Press.

Hallowell, A.I. 1955. *Culture and Experience*. Philadelphia: University of Pennsylvania Press.

Haraway, D. 1988. 'Situated Knowledges: The Science Question in Feminism and the Privilege of Partial Perspective', *Feminist Studies* 14(3): 575–600.

———. 1993. 'The Biopolitics of Postmodern Bodies: Determinations of Self in Immune System Discourse', in S. Lindenbaum and M. Lock (eds), *Knowledge, Power and Practice: The Anthropology of Medicine and Everyday Life*. Berkeley: University of California Press, pp. 364–411.

Harborne, J.B. and C.A. Williams. 2000. 'Advances in Flavonoid Research since 1992', *Journal of Phytochemistry* 55(6): 481–504.

Harris, L. 2002. 'An Evaluation of the Bronchodilator Properties of *Mentha longifolia* and *Artemisia afra*, Traditional Medicinal Plants Used in the Western Cape', master's thesis. Bellville: University of the Western Cape.

Heidegger, M. 1962. *Being and Time*. New York: Harper Perennial.

Heisenberg, W. 1971. *Physics and Beyond: Encounters and Conversations*. New York: Harper & Row.

Herzfeld, M. 2010. 'Senses', in A.C.G.M. Robben and J.A. Sluka (eds), *Ethnographic Fieldwork: An Anthropological Reader*. Malden, MA: Blackwell, pp. 431–42.

Hill, A.B. 1952. 'The clinical trial', *New England Journal of Medicine* 247: 113–19.

Hours, B. 1998. *L'idéologie humanitaire ou le spectacle de l'altérité perdue*. Paris: L'Harmattan.

Howes, D. 1990. 'Les techniques des sens', *Anthropologie et Sociétés* 14(2): 99–116.

———. 1991. 'Introduction: To Summon All the Senses', in D. Howes (ed.), *The Varieties of Sensory Experience: A Sourcebook in the Anthropology of the Senses*. Toronto: Toronto University Press, pp. 3–21.

Hsu, E. 2010. 'Introduction: Plants in Medical Practice and Common Sense—on the Interface of Ethnobotany and Medical Anthropology', in E. Hsu and S. Harris (eds), *Plants, Health and Healing: On the Interface of Ethnobotany and Medical Anthropology*. Oxford and New York: Berghahn Books, pp. 1–48.

———. 2012. 'Medical Anthropology in Europe: Quo Vadis?', *Anthropology and Medicine* 19(1): 51–61.

Hsu, E. and C. Low. 2008. *Wind, Life, Health: Anthropological and Historical Perspectives*. Malden, MA: Blackwell.

Husserl, E. 1950. *Idées directrices pour une phénoménologie*, trans. P. Ricoeur. Paris: Gallimard.

———. (1910–11) 2006. *The Basic Problems of Phenomenology: From the Lectures, Winter Semester, 1910–1911*, trans. I. Farin and J.G. Hart. Dordrecht: Springer.

Hutchings, A., et al. 1996. *Zulu Medicinal Plants: An Inventory.* Pietermaritzburg: University of Natal Press.

Indigenous Knowledge Systems Policy. 2004. Retrieved 10 December 2013 from http://www.wipo.int/tk/en/databases/creative_heritage/policy/link0007.html.

Ingold, T. 2000. *The Perception of the Environment: Essays on Livelihood, Dwelling and Skill.* London and New York: Routledge.

———. 2006. 'Rethinking the Animate, Re-animating Thought', *Ethnos* 71(1): 9–20.

———. 2008. 'Earth, Sky, Wind and Weather', in E. Hsu and C. Low (eds), *Wind, Life, Health: Anthropological and Historical Perspectives.* Malden, MA: Blackwell, pp. 17–36.

———. 2011. *Being Alive: Essays on Movement, Knowledge and Description.* London and New York: Routledge.

———. 2013. *Making: Anthropology, Archaeology, Art and Architecture.* London and New York: Routledge.

Ingold, T. and Hallam. 2007. 'Creativity and Cultural Improvisation: An Introduction', in E. Hallam and T. Ingold (eds), *Creativity and Cultural Improvisation.* New York: Berg, pp. 1–41.

Ives, S. 2014. 'Uprooting "Indigeneity" in South Africa's Western Cape: The Plant That Moves', *American Anthropologist* 116(2): 310–23.

Iwu, M.M. 1993. *Handbook of African Medicinal Plants.* Boca Raton, FL: CRC Press.

Jackson, M. (ed.). 1996. *Things as They Are: New Directions in Phenomenological Anthropology.* Bloomington: Indiana University Press.

Jackson, W.P.U. 1990. *Origins and Meanings of Names of South African Plant Genera.* Cape Town: University of Cape Town.

Jaeger, W. 1945. *Paideia: The Ideals of Greek Culture*, trans. G. Highet, 2nd ed., vol. 1. New York: Oxford University Press.

Janzen, J.M. 1992. *Ngoma: Discourses in Healing.* Berkeley and Los Angeles: University of California Press.

———. 2000. 'Afterword', in R. van Dijk, R. Reis and M. Spierenburg (eds), *The Quest for Fruition Through Ngoma: Political Aspects of Healing in Southern Africa.* Oxford: James Currey, pp. 155–68.

Javu, M. 2011. 'IKS Workshop on Intellectual Property Rights and Research on Traditional Medicines', Pretoria, Gauteng. Retrieved 27 August 2014 from http://www.nstf.org.za/ShowProperty?nodePath=/NSTF%20Repository/NSTF/files/Workshops/2011/rootsback.pdf.

Johnson, Q. 2011. *Phytomedicines: From Translation of IKS to Innovation for the Bioeconomy: briefing by the International Centre for Innovation Partnership in Science Phytomedicines.* Retrieved 28 August 2014 from http://www.pmg.org.za/report/20110907-prof-quinton-johnson-international-centre-innovation-partnerships-sci.

Johnson, Q., et al. 2007. 'A Randomized, Double Blind, Placebo-Controlled Trial of *Lessertia frutescens* in Healthy Adults', *PLOS Clinical Trials* 2(4): e16. doi:10.1371/journal. pctr.00220016.

Kaptchuk, T. 1998. 'Powerful Placebo: The Dark Side of the Randomized Controlled Trial', *Lancet* 351: 1722–25.

———. 2001. 'The Double-Blind, Randomized, Placebo-Controlled Trial: Gold Standard or Golden Calf?', *Journal of Clinical Epidemiology* 54: 541–49.

———. 2011. 'Placebo Studies and Ritual Theory: A Comparative Analysis of Navajo, Acupuncture and Biomedical Healing', *Philosophical Transactions of the Royal Society B* 366: 1849–58.

Katz, J. and T. Csordas. 2003. 'Phenomenological ethnography in sociology and anthropology', *Ethnography* 4(3) : 275–88.

Kay, L. E. 1993. *The Molecular Vision of Life: Caltech, the Rockefeller Foundation, and the Rise of the New Biology.* New York: Oxford University Press.

Kepe, T. 2003. '*Cannabis sativa* and Rural Livelihoods in South Africa: Politics of Cultivation, Trade and Value in Pondoland', *Development Southern Africa* 20(5): 605–15.

Kleinman, A. 1980. *Patients and Healers in the Context of Culture: An Exploration of the borderland Between Anthropology, Medicine and Psychiatry.* Berkeley: University of California Press.

———. 1988. *The Illness Narratives: Suffering, Healing, and the Human Condition.* New York: Basic Books.

———. 1999. 'Moral Experience and Ethical Reflection: Can Ethnography Reconcile Them? A Quandary for "The New Bioethics"', *Daedalus* 128(4): 69–97.

Kleinman, A., et al. (eds). 1976. *Medicine in Chinese Cultures: Comparative Studies of Health Care in Chinese and Other Societies.* Washington, D.C.: U.S. Government Printing Office for Fogarty International Center, NIH.

Knibbe, K. and P. Versteeg. 2008. 'Religion and Experience Assessing Phenomenology in Anthropology: Lessons from the Study of Religion and Experience', *Critique of Anthropology* 28: 47.

Knorr-Cetina, K. 1999. *Epistemic Cultures: How the Sciences Make Knowledge.* Cambridge, MA: Harvard University Press.

Kornegay, F. and C. Landsberg. 1998. 'Mayivuke ¡Africa! Can South Africa lead an African Renaissance?', *Policy: Issues and Actors* 11(1): 4.

Kroll, F. 2006. 'Roots and Culture: Rasta Bushdoctors of the Cape, SA', in W. Zips (ed.), *A Universal Philosophy: Rastafari in the Third Millennium.* Kingston: Ian Randle Publishers, pp. 215–55.

Küchler, S. 2006. 'Des matières qui travaillent: Une leçon pour l'anthropologie', *Anthropologie et sociétés* 30(3): 125–38.

Kusenbath, M. 2003. 'Street Phenomenology: The Go-along as Ethnographic Research Tool', *Ethnography* 4(3): 455–85.

Landsberg, C. and D. Hlophe. 1999. *The African Renaissance as a Modern South African Foreign Policy.* Paris: CERI.

Laplante, J. 2003. 'Le médicament aux frontières des savoirs humanitaires et autochtones', *Anthropologie et Sociétés* 27(2): 59–76.

———. 2004. *Pouvoir guérir: Médecines autochtones et humanitaires*. Québec: Presses de l'Université Laval.

———. 2006. 'Médicaments et médecines traditionnelles: Le cas d'interventions en santé internationale auprès des autochtones de l'Amazonie brésilienne', *Éthique Publique* 8(2): 143–51.

———. 2007. 'Trajectoire de savoirs entourant les médicaments', *Revue Internationale sur le Médicament* 1: 78–101.

———. 2009a. 'Plantes médicinales, savoirs et société: Vue des rastafaris sud-africains', *Drogues, santé et sociétés* 8(1): 93–121.

———. 2009b. 'South African Roots towards Global Knowledge: Music or Molecules?', *Anthropology Southern Africa* 32(1–2): 8–17.

———. 2012. '"Art de dire" Rastafari: Dagga et créativité musicale dans les townships sud-africains', *Drogues, santé et sociétés* 11(1): 90–106.

———. Forthcoming. 'Medicine Multiple: Ontologies in the Pre-clinical Trial of a South African Indigenous Medicine', in V. Hörbst, R. Gerrets and P. Schirripa (eds), *Diversifying Medical Pluralism. Medical Anthropological Perspectives on Plurality*. London: Palgrave Macmillan.

Laplante, J. and J. Bruneau. 2003. 'Aperçu d'une anthropologie du vaccin: Regards sur l'éthique d'une pratique humanitaire', *História, Ciências, Saúde-Manguinhos* 10(2): 519–38.

———. 2011. 'Streetwise Vaccination? Bio-politics Downtown Montreal', *Asian Journal of Canadian Studies* 17(1): 127–58.

Laplantine, F. 1992. *Anthropologie de la maladie*. Paris: Bibliothèque Scientifique Payot.

Lash, S. 2003. 'Reflexivity as Non-linearity', *Theory, Culture & Society* 20(2): 49–57.

Last, M. 1981. 'The Importance of Knowing about Not Knowing', *Social Science and Medicine* 15(3): 387–92.

Latour, B. 1987. *Science in Action: How to Follow Scientists and Engineers Through Society*. Cambridge, MA: Harvard University Press.

———. 2000. 'When Things Strike Back: A Possible Contribution of "Science Studies" to the Social Sciences', *British Journal of Sociology* 51(1): 107–23.

———. 2005. *Reassembling the Social: An Introduction to Actor-Network-Theory*. New York: Oxford University Press.

Latour, B. and S. Woolgar. 1988. *La vie de Laboratoire: La production des faits scientifiques*. Paris: La Découverte.

Lawn, S.D. and A.I. Zumla. 2011. 'Tuberculosis', *Lancet* 378: 57–72.

Leibing, A. 2004. *Tecnologias do corpo: Uma antropologia das medicinas no Brasil*. Rio de Janeiro: NAU Editora.

Leibing, A. and V. Tournay (eds). 2010. *Les technologies de l'espoir: La fabrique d'une histoire à accomplir*. Québec: Presses de l'Université Laval.

Leslie, C. 1976. *Asian Medical Systems*. Berkeley: University of California Press.

———. 1980. 'Medical Pluralism in World Perspective', *Social Science & Medicine B* 14(4): 191–95.

Lévi-Strauss, C. 1949. 'L'efficacité symbolique', *Revue de l'histoire des religions* 135(1): 5–27.

———. 1962. *La pensée sauvage*. Paris: Plon.

———. 1971. *Mythologiques: L'homme nu*. Paris: Plon.

Littlewood, R. (ed.). 2007. *On Knowing and Not Knowing in the Anthropology of Medicine*. Walnut Creek, CA: Left Coast Press.

Liu, C.-Z., H.-Y. Zhou and Y. Zhao. 2007. 'An Effective Method for Fast Determination of Artemisinin in *Artemisia annua* L.', *Analytica Chimica Acta* 581(2): 298–302.

Liu, N.Q., F. Van der Kooy and R. Verpoorte. 2009. '*Artemisia afra*: A Potential Flagship for African Medicinal Plants?', *South African Journal of Botany* 75(2): 185–95.

Lock, M. 2005. 'Eclipse of the Gene and the Return of Divination', *Current Anthropology* 46: S47–S60.

Lock, M. and J. Faqurhar (eds). 2007. *Beyond the Body Proper: Reading the Anthropology of Material Life*. Durham, NC: Duke University Press.

Lock, M. and V.-K. Nguyen. 2010. *An Anthropology of Biomedicine*. West Sussex: Wiley-Blackwell.

Lock, M. and M. Nichter. 2002. 'Introduction', in M. Nichter and M. Lock (eds), *New Horizons in Medical Anthropology: Essays in Honour of Charles Leslie*. London: Routledge, pp. 1–34.

Low, C. 2008. 'Khoisan Wind: Hunting and Healing', in E. Hsu and C. Low (eds), *Wind, Life, Health: Anthropological and Historical Perspectives*. Oxford: Blackwell, pp. 65–84.

Löwy, I. 2000. 'Trustworthy Knowledge and Desperate Patients: Clinical Tests for New Drugs from Cancer to AIDS', in M. Lock, A. Young and A. Cambrosio (eds), *Living and Working with the New Medical Technologies: Intersections of Inquiry*. Cambridge: Cambridge University Press, pp. 41–81.

Lynch, M.E. 1988. 'Sacrifice and the Transformation of the Animal Body into a Scientific Object: Laboratory Culture and Ritual Practice in the Neurosciences', *Social Studies of Science* 18: 265–89.

Mandela, N. 1994. *Inaugural Speech. Statement of the President of the African National Congress, Pretoria*. Retrieved 19 December 2013 from http://www.africa.upenn.edu/ Articles_Gen/Inaugural_Speech_17984.html

Mander, M., et al. 2012. *Economics of the Traditional Medicine Trade in South Africa*. Retrieved 12 February 2012 from http://www.hst.org.za/uploads/files/chap13_07.pdf.

Mantula, R.G.S. 2006. 'Establishment of the Rastafari Forum at the North-West University'. Issued by the North West Roots Radics Research and International Liaison Desk, December.

Marks, H. 1987. 'Ideas as Reforms: Therapeutic Experiments and Medical Practice', Ph.D. dissertation. Cambridge, MA: Massachusetts Institute of Technology.

———. 1997. *The Progress of Experiment: Science and Therapeutic Reform in the United States, 1900–1990*. Cambridge: Cambridge University Press.

———. 1999. *La médecine des preuves: Histoire et anthropologie des essais cliniques (1900–1990)*. Cambridge: Cambridge University Press.

Marley, Z. 1997. 'Introduction', in G. Hausman (ed.), *The Kebra Nagast: The Lost Bible of Rastafarian Wisdom and Faith From Ethiopia and Jamaica*. New York: St. Martin's Press, pp. 7–11.

Marten, B. 1720. *A New Theory of Consumptions: More Especially of a Phthisis or Consumption of the Lungs*. London: T. Knaplock.

Matsabisa, M. 2008. 'MRC Clinical Trials on Indigenous Medicines', African Indigenous Medicine Symposium (AIMS): Clinical Trials, Cape Town, 25 August 2008.

Mauss, M. 1934. 'Les techniques du corps', *Journal de Psychologie* 32: 3–4.

Mavimbela, V. 1998. 'The African Renaissance: A Workable Dream', in G. le Pere, A. van Nieuwkerk and K. Lambrechts (eds), *South Africa and Africa: Reflections on the African Renaissance*, Johannesburg: Foundation for Global Dialogue, Occasional Paper 17.

McLuhan, M. 1971. *La galaxie Gutenbert*. Montréal: Hurtubise HMH.

———. 1972 *Pour comprendre les média : les prolongements technologiques de l'homme*. Montréal: Hurtubise HMH.

Mead, M. 1985. 'Coming of age: Margaret Mead' Video. Directed by André Singer. Great Britain: Strangers Abroad.

Merleau-Ponty, M. 1945. *La phénoménologie de la perception*. Paris: Gallimard.

———. 1962. *Phenomenology of Perception*, trans. J. Edie. Evanston, IL: Northwestern University Press.

———. 1964. *L'œil et l'Esprit*. Paris: Gallimard.

———. 1995. *La Nature. Notes de cours du Collège de France*. Paris: Seuil.

———. 2012. *Phenomenology of Perception*. Abingdon: Routledge.

Miller, R.N. 1972. 'The Positive Effect of Prayer on Plants', *Psychic* 3(5): 24–25.

Mol, A. 2002. *The Body Multiple: Ontology in Medical Practice*. Durham, NC: Duke University Press.

Moridani, M.Y., G. Galati and P.J. O'Brien. 2002. 'Comparative Quantitative Structure Toxicity Relationships for Flavonoids Evaluated in Isolated Rat Hepatocytes and HeLa Tumor Cells', *Chemico-Biological Interactions* 139: 251–64.

Mthembu, N. 2007. 'Contemporary Changes in South Africa (Azania): Rastafari Community Experience at Ethekwini', Educational Talk: Why Rastafari in Africa, South Africa, 17–23 February 2007.

Muganga, R. 2005. 'Luteolin Levels in Selected Folkloric Preparations and the Bioavailability of Luteolin from *Artemisia afra* Aqueous Extract in the Vervet Monkey', master's thesis. Bellville: University of the Western Cape.

Mukinda, J.T. 2005. 'Acute and Chronic Toxicity of the Flavonoid-Containing Plant, *Artemisia afra* in Rodents', master's thesis. Bellville: University of the Western Cape.

Mukinda, J. and J. Syce. 2007. 'Acute and Chronic Toxicity of the Aqueous Extract of *Artemisia afra* in Rodents', *Journal of Ethnopharmacology* 112: 138–44.

National Institutes of Health (NIH). 2012a. 'Almanac'. Retrieved 17 November 2013 from http://www.nih.gov/about/almanac/organization/NCCAM.htm#events.

————. 2012b. 'History'. Retrieved 17 November 2013 from http://www.history
.nih.gov/exhibits/history/docs/page_06.html.

Ndaki, K. 2005. 'The Good Doctors', *Fairlady* 72 (February).

Neuwinger, H.D. 2000. *African Traditional Medicine: A Dictionary of Plant Use
and Applications.* Stuttgart: Medpharm Scientific.

Nguyen, V.-K. 2004. 'Antiretroviral Globalism, Biopolitics and therapeutic Cit-
izenship', in A. Ong and S. Collier (eds), *Global Assemblages: Technology,
Politics and Ethics.* London: Blackwell, pp. 124–44.

Niaah, J.A. 2003. 'Poverty (Lab) Oratory: Rastafari and Cultural Studies', *Cultural Studies* 17(6): 823–42.

Nichter, M. 1992. 'Ethnomedicine: Diverse Trends, Common Linkages', in M.
Nichter (ed.), *Anthropological Approaches to the Study of Ethnomedicine.*
Amsterdam: Gordon and Breach, pp. 223–59.

Nichter, M. and M. Lock (eds). 2002. *New Horizons in Medical Anthropology:
Essays in Honour of Charles Leslie.* London: Routledge, pp. 1–34.

Ntutela, S., et al. 2009. 'Efficacy of *Artemisia afra* Phytotherapy in Experimental
Tuberculosis', *Tuberculosis* 1: S33–S40.

Olivier, L.E. 2011. 'Rastafari Bushdoctors and the Challenges of Transforming
Nature Conservation in the Boland Area', masters' thesis, Stellenbosch: The
University of Stellenbosch.

O'Neil, J. 1985. *Five bodies: The shape of modern society.* Ithaca, NY: Cornell
University Press.

Orbinski, J. 1999. 'Nobel Lecture', Médecins Sans Frontières, Oslo, 10 December. Retrieved 17 December 2013 from http://www.nobelprize.org/
nobel_prizes/peace/laureates/1999/msf-lecture.html.

Osseo-Asare, A.D. 2014. *Bitter Roots: The Search for Healing Plants in Africa.*
Chicago and London: The University of Chicago Press.

Ostenfeld-Rosenthal, A.M. 2012. 'Energy Healing and the Placebo Effect: An
Anthropological Perspective on the Placebo Effect', *Anthropology and Medicine* 19(3): 327–38. doi:10.1080/13648470.2011.646943.

Parkins, D. 2008. 'Wafting on the Wind: Smell and the Cycle of Spirit and
Matter', in E. Hsu and C. Low (eds), *Wind, Life, Health: Anthropological and
Historical Perspectives.* Malden, MA: Blackwell, pp. 37–50.

Patil, G.V., S.K. Dass and R. Chandra. 2011. '*Artemisia afra* and Modern Diseases', *Pharmacogenomics & Pharmacoproteomics* 2(3): 1–22.

Patwardhan, B. and R.A. Mashelkar. 2009. 'Traditional Medicine-Inspired Approaches to Drug Discovery: Can Ayurveda Show the Way Forward?', *Drug
Discovery Today* 14(15–16): 804–11.

Paul, B.D. 1955. *Health, Culture and Community: Case Studies of Public Reactions to Health Programs.* New York: Russell Sage Foundation.

Pauly, P.J. 1987. *Controlling Life: Jacques Loeb and the Engineering Ideal in
Biology.* New York: Oxford University Press.

Pelican, M. 2009. 'Complexities of Indigeneity and Autochthony: An African
Example', *American Ethnologist* 36(1): 52–65.

Petryna, A. 2005. 'Ethical Variability: Drug Development and Globalizing Clinical Trials', *American Ethnologist* 32(2): 183–97.

———. 2007a. 'Clinical Trial Offshored: On Private Sector Science and Private Health', *Biosocieties* 2: 21–40.

———. 2007b. 'Experimentality: On the Global Mobility and Regulation of Human Subjects Research', *Polar: Political and Legal Anthropology Review* 30: 288–304.

———. 2009. *When Experiments Travel: Clinical Trials and the Global Search for Human Subjects*. Princeton, NJ: Princeton University Press.

Petryna, A., A. Lakoff and A. Kleinman. 2006. *Global Pharmaceuticals: Ethics, Markets, Practices*. Durham, NC: Duke University Press.

Pickering, J.V. 2000. *Ways of Knowing: A New History of Science Technology and Medicine*. Manchester: Manchester University Press.

Philander, L. 2010. 'An Emergent Ethnomedicine: Rastafari Bush Doctors in the Western Cape, South Africa', Ph.D. thesis, Tucson: The University of Arizona.

Phillips, M. 2006. 'Persuading Africans to Take Their Herbs with Some Antivirals', *Wall Street Journal*, 5 May.

Pollan, M. 2002. *Botany of Desire: A Plant's-Eye View of the World*. New York: Random House.

Pool, R. 1994. 'On the Creation and Dissolution of Ethnomedical Systems in the Medical Ethnography of Africa', *Africa* 64 (1): 1–20.

Prins, G. 1989. 'But What Was the Disease? The Present State of Health and Healing in African Studies', *Past and Present* 124: 159–79.

Rabinow, P. 1996. *Essays on the Anthropology of Reason*. Princeton, NJ: Princeton University Press.

Reid, A.M. 2014. 'Rastas on the Road to Healing: Plant-Human Mobilities in Cape Town, South Africa', master's minor dissertation, Cape Town: University of Cape Town.

Revicki, D.A. and L. Frank. 1999. 'Pharmacoeconomic Evaluation in the Real World: Effectiveness Versus Efficacy Studies', *Pharmacoeconomics* 15(5): 423–34.

Rheinberger, H.J. 1995. 'Beyond Nature and Culture: A Note on Medicine in the Age of Molecular Biology', *Science in Context* 8(1): 249–63.

Rivers, W.H.R. 1901. 'Introduction and Vision', in A.C. Haddot (ed.), *Report of the Cambridge Expedition to the Torres Straits*. Cambridge: The University Press, pp. 1–5.

———. 1924. *Medicine, Magic and Religion: The FitzPatrick Lectures delivered before The Royal College of Physicians of London in 1915 and 1916*. London: Kegan Paul, Trench, Trubner & Co.

Roberts, M. 1990. *Indigenous Healing Plants: South Africa*. Johannesburg: Southern Book Publishers, pp. 226–28.

Rood, B. 1994. *Uit die veldapteek*. Cape Town: Tafelberg.

Rorty, R. 1979. *Philosophy and the Mirror of Nature*. Princeton, NJ: Princeton University Press.

Rose, N. 2007. *The Politics of Life Itself: Biomedicine, Power and Subjectivity in the Twenty-First Century*. Princeton, NJ: Princeton University Press.

Ross, F.C. 2010. *Raw Life, New Hope: Decency, Housing and Everyday Life in a Post-apartheid Community*. Cape Town: UCT Press.

Roy, B. 2002. *Sang sucré, pouvoirs codés, médecine amère: Diabète et processus de construction identitaire: les dimensions socio-politiques du diabète chez les Innus de Pessamit.* Québec: Les presses de l'Université Laval.

Roy, P. 1996. 'Variabilité dans l'usage thérapeutique des plantes chez les Totonaques', *Recherches Amérindiennes au Québec* 26(1): 43–54.

Rudd, P. 1979. 'In Search of the Gold Standard for Compliance Measurement', *Archives of Internal Medicine* 139: 627–28.

Saethre, E.J. and J. Stadler. 2010. 'Gelling Medical Knowledge: Innovative Pharmaceuticals, Experience, and Perceptions of Efficacy', *Anthropology & Medicine* 17(1): 99–111.

Samudra, J.K. 2008. 'Memory in Our Body: Thick Participation and the Translation of Kinesthetic Experience', *American Ethnologist* 35(4): 665–81.

Scheper-Hughes, N. and M. Lock. 1987. 'The Mindful Body: A Prolegomenon to Future Work in Medical Anthropology', *Medical Anthropology Quarterly* 1(1): 6–41.

Schiller, F. 1982. *On the Aesthetics and Education of Man,* ed. and trans. E.M. Wilkinson and L.A. Willoughby. Oxford: Clarendon Press.

Schultes, R.E. and S. von Reis. 1995. *Ethnobotany: Evolution of a Discipline.* Portland: Dioscorides Press.

Scott, C. 1989. 'Knowledge Construction among Cree Hunters: Metaphors and Literal Understanding', *Journal de la Société des Américanistes* 75: 193–208.

Shapiro, K. 1987. 'Doctors or Medical Aids: The Debate over the Training of Black Medical Personnel for the Rural Black Population in South Africa in the 1920s and 1930s', *Journal of Southern African Studies* 13(2): 55–75.

Shimoi, K., et al. 1998. 'Intestinal Absorption of Luteolin and Luteolin 7-O-ßglucoside in Rats and Humans', *FEBS Letters* 438: 220–24.

Skibola, C.F. and M.T. Smith. 2000. 'Potential Health Impacts of Excessive Flavonoid Intake', *Free Radical Biology & Medicine* 29: 375–83.

Smart, T. 2005. Traditional healers being integrated into HIV care and treatment in Kwazulu-Natal. The nam aidsmap website, retrieved 25 August 2014 from http://www.aidsmap.com/Traditional-healers-being-integrated-into-HIV-care-and-treatment-in-Kwazulu-Natal/page/1421024/.

Smith, N.H., et al. 2009. 'Myths and Misconceptions: The Origin and Evolution of *Mycobacterium tuberculosis*', *Nature Reviews Microbiology* 7: 537–44.

Star, S.L. 1989. *Regions of the Mind: Brain Research and the Quest for Scientific Certainty.* Stanford, CA: Stanford University Press.

Star, S.L. and M. Lampland. 2009. *Standards and Their Stories: How Quantifying, Classifying, and Formalizing Practices Shape Everyday Life.* Ithaca, NY: Cornell University Press.

Stolberg, H.O., G. Norman and I. Trop. 2004. 'Randomized Controlled Trials', *American Journal of Roentgenology* 183: 1539–44.

Stoller, P. 1986. *The Taste of Ethnographic Things: The Senses in Anthropology.* Philadelphia: University of Pennsylvania Press.

Streptomycin in Tuberculosis Trials Committee. 1948. 'Streptomycin Treatment of Pulmonary Tuberculosis: A Medical Research Council Investigation', *British Medical Journal* 2 (4582): 769–82.

Stroeken, K. 2008. 'Sensory Shifts and "Synaesthetics" in Sukuma Healing', *Ethnos* 73(4): 466–84.

———. 2012. 'Health Care Decisions by Sukuma "Peasant Intellectuals": A Case of Racial Empiricism?', *Anthropology & Medicine* 19(1): 119–28.

Sunder Rajan, K. 2006. *Biocapital: The Constitution of Postgenomic Life*. Durham, NC: Duke University Press.

Tart, C.T. 1971. *On Being Stoned, a Psychological Study of Marihuana Intoxication*. Palo Alto, CA: Science and Behaviour Books.

The International Center for Indigenous Phytotherapy Studies (TICIPS). 2005. 'Statement'. Retrieved 10 December 2013 from http://www.wlbcenter.org/ticips.htm.

Thompson, G. 2003. 'Bushmen Squeeze Money from a Humble Cactus', *New York Times*, 1 April.

Thompson, R.F. 1983. *Flash of the Spirit*. New York: Random House.

Thring, T.S.A. and F.M. Weitz. 2006. 'Medicinal Plant Use in the Bredasdorp/Elim Region of the Southern Overberg in the Western Cape Province of South Africa', *Journal of Ethnopharmacology* 103: 261–75.

Timmermans, S. and M. Berg. 2003. *The Gold Standard: The Challenge of Evidence-Based Medicine and Standardization in Health Care*. Philadelphia: Temple University Press.

Tompkins, P. and C. Bird. 1972. 'Love Among Cabbages: Sense and Sensibility in the Realm of Plants', *Harper's Magazine*, November, 90–6.

Turner, B.S. 1992. *Regulating Bodies: Essays in Medical Sociology*. London: Routledge.

Turner, V. 1968. *Drums of Affliction: A Study of Religious Processes Among the Ndembu of Zambia*. Oxford: Oxford University Press.

———. 1977. 'Variations On a Theme of Liminality', in S.F. Moore and B.G. Myerhoff (eds), *Secular Ritual*. Assen, Netherlands: Van Gorcum, pp. 36–52.

Twyman, R. 2004. 'A Brief History of Clinical Trials', The Human Genome website. Retrieved 10 December 2013 from http://genome.wellcome.ac.uk/doc_WTD020948.html.

Tylor, S.A. 1986. 'Post-modern Ethnography: From Document of the Occult to Occult Document', in J. Clifford and G.E. Marcus (eds), *Writing Culture: The Poetics and Politics of Ethnography*. Berkeley: University of California Press, pp. 122–40.

Van Beek, W.E. and P.M. Banga. 1992. 'The Dogon and their trees' in E. Croll and D. Parkin (eds), *Bush Base: Forest Farm. Culture, Environment and Development*. London: Routledge.

van der Geest, S. and S.R. Whyte. 1988. *The Context of Medicines in Developing Countries: Studies in Pharmaceutical Anthropology*. Dordrecht: Kluwer Academic.

van der Geest, S., S.R. Whyte and A. Hardon. 1996. 'The Anthropology of Pharmaceuticals: A Biographical Approach', *American Review of Anthropology* 25: 153–78.

van Wyk, A. 2005. 'Evaluation of Guidelines for Clinical Trials of Traditional Plant Medicines', master's thesis. Bellville: University of the Western Cape.

van Wyk, B.-E. and N. Gericke (eds). 2007. *People's Plants: A Guide to Useful Plants of Southern Africa*. Pretoria: Briza.

van Wyk, B.-E., B. van Oudtshoorn and N. Gericke. 1997. *Medicinal Plants of South Africa*, 1st ed. Pretoria: Briza.

———. 2000. *Medicinal Plants of South Africa*, 2nd ed. Singapore: Tien Wah Press.

Vaughan, M. 1991. *Curing Their Ills: Colonial Power and the African Illness*. Cambridge: Polity Press.

Verhoef, M.J., A.L. Casebeer and R.J. Hilsden. 2002. 'Assessing Efficacy of Complementary Medicine: Adding Qualitative Research Methods to the "Gold Standard"', *The Journal of Alternative and Complementary Medicine* 8(3): 275–81.

Verhoef, M.J., et al. 2005. 'Complementary and Alternative Medicine Whole Systems Research: Beyond Identification of Inadequacies of the RCT', *Complementary Therapies in Medicine* 13: 206–12.

Vigarello, G. 2007. 'La vision du corps dans les sciences sociales', in M. Wieviorka (ed.), *Les sciences sociales en mutation*. Paris: Éditions sciences humaines, pp. 83–90.

Vilakazi, H.W. 2006. 'The Coming Revolution in Modern Medicine: Dreaded Diseases and African Traditional Medicine – a Proposal'. Meeting with President Thabo Mbeki, Pretoria, 10 May.

von Koenen, E. 2001. *Medicinal, Poisonous and Edible Plants in Namibia*. Windhoek and Göttingen: Klaus Hess Verlag.

von Uexküll, J. 1937. 'The New Concept of Umwelt: A Link between Science and the Humanities', trans. G. Brunow, *Die Eriehung* 13(5): 185–99.

Vuckovic, N. 2002. 'Integrating Qualitative Methods in Randomized Controlled Trials: The Experience of the Oregon Center for Complementary and Alternative Medicine', *The Journal of Alternative and Complementary Medicine* 8(3): 225–27.

Wahlberg, A. 2006. 'Bio-politics and the Promotion of Traditional Herbal Medicine in Vietnam', *Health: An Interdisciplinary Journal for the Social Study of Health, Illness and Medicine* 10(2): 123–47.

Waithaka, J. 2004. 'The Evaluation of Markers for Quality Control Studies of Flavonoid-Containing Medicinal Preparations', master's thesis. Bellville: University of the Western Cape.

Waldram, J.B. 2000. 'The Efficacy of Traditional Medicine: Current Theoretical and Methodological Issues', *Medical Anthroplogy Quarterly* 14(4): 603–25.

Watt, J.M. and M.G. Breyer-Brandwijk. 1962. *The Medicinal and Poisonous Plants of Southern and Eastern Africa*, 2nd ed. London: Livingstone.

Wattchow, B. 2012. 'Ecopoetic Practice: Writing the Wounded Land', *Cultural Studies – Critical Methodologies* 12(1): 15–21.

Whyte, S.R., S. van der Geest and A. Hardon. 2002. *Social Lives of Medicines*. Cambridge: Cambridge University Press.

World Health Organization. 2008. 'Fact Sheet No. 134', December. Retrieved 12 December 2013 from http://www.siav-itvas.org/images/stories/doc/ago puntura_scientifica/WHO_Traditional_medicine_2008.pdf.

Wreford, J. 2008. *Working with Spirit: Experiencing Izangoma Healing in Contemporary South Africa*. London and New York: Berghahn Books.

Yawney, C.D. 1978. 'Lions in Babylon: The Rastafarians of Jamaica as Visionary Movement', Ph.D. dissertation. Montreal: McGill University.

Yu, S.-D. 2009. 'Introduction: Bodily Cultivation as a Mode of Learning', *Taiwan Journal of Anthropology* 7(2): 3–12.

Index

A

aesthetics, 61, 63, 241–42, 248–57
African Renaissance, 148–49, 169, 182, 235–41
Afrikaans, 84, 91, 160–61, 168–69
Afrikaner, 168, 188, 217
Agamben, Giorgio, 67, 227n1
ancestors, 28, 36, 69, 99, 109, 122–29, 149–54, 156, 177, 184–86, 191–200, 242
animal life, 65–66, 71, 74, 177–78, 181, 228n2, 250
    in *muthi*, 99, 110, 149, 174, 191, 195–97
    sacrifice, 36, 109, 152, 194–96
anthropology in life and medicine, 1, 45, 197, 213, 230, 241, 248–57
anthropology of the senses, 45–46, 62–65
apartheid, 11n8, 84, 105–6, 111, 168–69, 188–89, 217, 241
    postapartheid, 28, 157, 164, 168–69, 171, 182, 226, 234
Arendt, Hannah, 38, 61
Artemisia afra (*A. afra*) trails and trials, 6, 18, 66–67, 76–77, 103–4, 111–34, 142, 201, 210, 214, 221, 229–48
    botanical inventories, 13, 18–21, 31, 45, 107, 118, 135, 147, 204
    botanical plant entity, 2, 5, 18–21, 31–33, 36–38, 80, 98, 102, 121, 245
    cultivation, 27–29, 100, 103, 125, 177, 217

kin, 172–75, 179, 181, 213, 215
    preclinical study in tuberculosis, 2–7, 26–43, 143–45, 182, 206–8, 212–14
    'standard' plant, 112
    synergies, 16, 27, 43n6, 214, 221
    toxicity studies, 20–21, 26–27, 33–34, 83, 116–17
    variability, 27, 29, 31, 37, 71, 92, 101, 103, 219
    See also mediums; *umhlonyane*
Artemisia annua (*A. annua*), 3, 43n8

B

becomings, 1–6, 22, 28, 53, 58, 64, 70, 75n1, 76, 105, 111, 123, 127–28, 150, 177–78, 185–87, 191–95, 198, 200, 221, 230–31, 244, 252–53
    lines of becoming, 4, 9, 17–18, 30, 32, 41, 44, 50, 54, 60, 75n4, 77, 151, 175, 204, 207, 229–31, 233, 244, 248, 256
    trails of becoming, 53, 67
    worlds of becoming, 45, 48–49, 67–75, 111
biochemistry. *See under* molecular biology
biodiversity, 102, 110, 238–39, 256
biological bodies, 1, 16, 24–25, 38, 48–49, 51, 73, 103, 117, 132n7, 133, 139, 153–56, 210, 217, 229

276

www.ingramcontent.com/pod-product-compliance
Lightning Source LLC
Chambersburg PA
CBHW070912030426
42336CB00014BA/2377